Quantum Mechanics and Avant-Garde Music

Rakhat-Bi Abdyssagin

Quantum Mechanics and Avant-Garde Music

Shadows of the Void

 Springer

Rakhat-Bi Abdyssagin
Almaty, Kazakhstan

Former Visiting Scholar at Wolfson College
University of Oxford
Oxford, UK

Institut für Musikwissenschaft
Goethe-Universität Frankfurt am Main
Frankfurt, Germany

Moth Quantum
London, UK

ISBN 978-3-031-63160-3   ISBN 978-3-031-63161-0   (eBook)
https://doi.org/10.1007/978-3-031-63161-0

This Springer imprint is published by the registered company Springer Nature Switzerland AG
The registered company address is: Gewerbestrasse 11, 6330 Cham, Switzerland

If disposing of this product, please recycle the paper.

Foreword

Exploring Quantum Harmonies: The Symphony of Quantum Music

In the vast realm of the implication of quantum mechanics lies a topic yet unexplored by many: quantum music. This unique book journeys into this domain by analysing the musical analogues of various quantum effects. Much more than that, Rakhat-Bi takes us on an incredible voyage involving a comparative study between the developments of quantum physics and avant-garde music.

As we delve into this realm, we encounter intriguing concepts and harmonious possibilities that stretch the boundaries of traditional views according to which science and arts develop independently of and without regard for each other. Much like the exotic behaviour of subatomic particles, this book challenges our perceptions and invites us to embrace a new paradigm for an important part of the intellectual history of the twentieth century. In this exploration, we shall venture through the key principles and enchanting melodies that comprise the symphony of quantum music. And Rakhat-Bi, himself a highly accomplished musician as well as an aficionado of quantum physics, is a perfect guide to take us on a wonderful journey exploring these ideas and much much more in this wonderful and passionate book.

At the heart of quantum physics lies the notion of superposition, a fundamental principle where states exist simultaneously in a blend of possibilities until measured. In the musical realm, a single note can exist in a multitude of frequencies, overlapping and intertwining to create intricate melodies that defy classical conventions. Each performance becomes a voyage into the unknown, as musicians harness the power of superposition to weave tapestries of sound that resonate with the very fabric of the quantum universe.

Can we hear the sound of a quantum superposition? No, this is not a Zen koan, like the clapping of one hand. It is a genuinely exciting question regarding the ability of human beings to perceive quantum effects. I was lucky to meet Rakhat-Bi in 2022 and learn from him about many parallels between the developments in modern music and modern physics in the twentieth century. Not only are there parallels of the intellectual kind but also there is the intriguing fact that many pioneers of quantum mechanics were, in fact, themselves proficient musicians.

Then there is quantum entanglement, a phenomenon that Einstein called 'a spooky action at a distance'. In the world of quantum entanglement, particles become interconnected in a phenomenon that transcends physical distance. Similarly, in the realm of quantum music, entangled melodies emerge as notes become intricately linked, resonating in perfect harmony regardless of spatial separation. As I've learnt from Rakhat-Bi, and to my great surprise, it is possible to play a number of notes on a wind instrument simultaneously whose overall sound cannot be written down as a note in the musical score! This is a perfect musical analogue to entanglement, since entanglement tells us that the total is more than the sum of its parts in the quantum world.

This entanglement of musical elements allows for compositions that transcend traditional notions of rhythm and melody, embracing a fluidity that mirrors many possibilities encased by the quantum world. Through entangled melodies, quantum musicians explore the boundless possibilities of sonic expression, crafting compositions that resonate with a profound sense of unity and coherence, challenging the views of both space and time.

Much like Heisenberg's Uncertainty Principle, which dictates the inherent limits of precision in measuring complementary properties of particles, the Uncertainty Principle of Sound governs the elusive nature of musical expression. In the realm of quantum music, rhythms fluctuate unpredictably, melodies shimmer with ambiguity, and harmonies dissolve into a kaleidoscope of sonic textures. Yet, it is within this uncertainty that the true beauty of quantum music emerges, inviting listeners to embrace the ephemeral nature of sound and surrender to the ever-changing symphony of the quantum universe.

One of the consequences of the uncertainty principle is the existence of a quantum vacuum. Even when there are no particles around, a quantum vacuum can have observable effects. In fact, one could even argue that the universe and everything it contains is actually nothing but 'the shadows of the quantum void'. In the same way, silence in music can communicate profound information to us as Rakhat-Bi convincingly argues throughout his undertaking.

I am sure that the reader of Rakhat-Bi's book will be left with a profound appreciation for the boundless creativity and infinite possibilities that this subject offers. As we read with open hearts and minds, we become participants in a cosmic dance of sound, where the truths of the quantum world converge with the timeless beauty of musical expression. And in this convergence, we find not only harmony but also a deeper understanding of the interconnectedness of all things.

Oxford, UK   Vlatko Vedral
April 2024

Acknowledgements

The acquisition of new knowledge and its application in all spheres of life is an integral part of human activity. Investigation of interconnections between science, music and art are extremely interesting and always relevant. My research on this topic started during my school years; in 2013, I published a short work *Mathematics and Contemporary Music*, and since that time I composed a number of music works inspired by science, as well as some scientific articles. The active phase of my research on correlations between quantum mechanics and avant-garde music was in 2021–2023, when I completed the initial manuscript of my book. The manuscript of my book greatly benefited from my discussions with numerous specialists. First of all, I express my deepest gratitude to physicist Vlatko Vedral—Professor of Quantum Information Science at the University of Oxford—who has been supporting my work with great interest and enthusiasm; I am also very thankful to physicist Chiara Marletto. Special thanks go to physicist Bob Coecke—Chief Scientist at Quantinuum—for providing great support for my book and scientific investigations.

The advice and opinions of distinguished physicists, Nobel Prize winners Brian Josephson (University of Cambridge), Tony Leggett (University of Illinois Urbana-Champaign) and David Politzer (Caltech) are exceedingly valuable to me, and I am immensely grateful to them.

During my research stay as a Visiting Scholar of Wolfson College, University of Oxford in October–December 2023, I was lucky enough to discuss the manuscript of my book with leading academics: physicists Julian Barbour,

Andrew Briggs, David Deutsch, Tim Palmer; philosophers Daniel Isaacson, Christopher G. Timpson; biologist Denis Noble; musicologists Philip Ross Bullock, Jonathan Cross, Daniel M. Grimley and others. Nobel laureates biologist Martin Chalfie (Columbia University) and biochemist Thomas C. Südhof (Stanford University) were kind to provide information about their relation to music. I highly appreciate our discussions with physicists Michael Berry (University of Bristol), John Ellis (King's College London), Raymond E. Goldstein (University of Cambridge), Antony Valentini; historian of science Cathryn Carson (UC Berkeley); philosopher Alex Carter (University of Cambridge).

Insights of Ian Stewart (University of Warwick) on the essence of mathematics and his advice on the manuscript were crucial for my work. I express profound gratitude to mathematician Richard Jozsa (University of Cambridge) for our discussions of quantum entanglement and for his advice regarding the structure of the manuscript. I thank philosophers of physics Harvey Brown (University of Oxford) and Jeremy Butterfield (University of Cambridge) for scrupulous reading the entire manuscript and providing extremely useful comments, philosopher Adrian W. Moore (University of Oxford) for our correspondence about Gödel's incompleteness theorems, philosopher Timothy Williamson (University of Oxford) for his helpful comments on the nature of mathematics, writer Georgina Ferry and composer Yuri Kasparov for our discussions of specific chapters of the book.

I have been tremendously lucky to study and learn the art of composition from the greatest composers of contemporary music, to whom I am eternally grateful. Beat Furrer opened the way for me to the world of new music by inviting me in 2012 to participate in the International Impuls Academy in Graz, Austria. And since the age of 14, I have been a regular participant of this wonderful festival. Impuls Graz is one of the world's largest international academies of composers and performers of contemporary music, where key processes in the evolution of contemporary music take place and a new face of the musical avant-garde is being formed. During Impuls Academies I have been able to learn and master the newest and the most advanced composing techniques, and gained knowledge in discussions with established composers. I am very grateful to M° Ivan Fedele, who was my supervisor during my postgraduate (doctoral) studies at Accademia Nazionale di Santa Cecilia in Rome, to M° Alessandro Solbiati, who was my supervisor during my postgraduate (doctoral) studies at Conservatorio G.Verdi di Milano, and also to the great composer Tristan Murail, for our discussions.

It was an exceptional honour and privilege to meet in person with children of the great physicist Werner Heisenberg. W.Heisenberg's son Martin

Heisenberg with his wife Apollonia, and W.Heisenberg's daughters Christine Mann and Barbara Blum provided me with a great deal of fascinating and rare information about the tremendous role that music played in the life of their father, as well as giving me access to unique and previously unpublished materials from their family archive. An especially fantastic honour for me was to play the Grand Piano of Werner Heisenberg which he bought in the 1930s for the money of the Nobel Prize. It was an enriching experience to meet with the members of the board of the Heisenberg Society in Munich, physicists Konrad Kleinknecht and Wolfgang Dünnweber.

Materials of the book have been discussed during my presentation at the scientific seminar in Vlatko Vedral's group working in quantum physics in Oxford (21st April 2023), and at the first Wolfson Quantum Foundations Discussion of the year, Wolfson College, University of Oxford (24th October 2023). I am grateful to musicologists Prof. Dr. Thomas Schipperges and Prof. Dr. Jörg Rothkamm for hosting my presentation at the Musikwissenschaftliches Institut, Eberhard Karls Universität Tübingen (31st October 2023). Some elements of my research have been discussed during my presentation at the Third Annual Meeting of Kazakh Physical Society, National Nuclear Center of the Republic of Kazakhstan, Kurchatov, Kazakhstan (8th June 2023).

I express deep gratitude to musicologist Prof. Dr. Magdalena Zorn (Goethe-Universität Frankfurt am Main) for our discussions on selected works of Karlheinz Stockhausen and John Cage, and for supporting my research.

I would also like to thank Dr. Zachary Evenson—Senior Editor at Springer Nature—for supporting this project and for his high professionalism.

April 2024   Rakhat-Bi Abdyssagin

Contents

List of Figures

Part I

Origins and Development of Avant-Garde Music in Faces

1

Prologue

The progress of thought and outlook finds its reflection in different fields of human activity. The development of science and the evolution of music represent the special states of the human mind. Peaks of such mind (mental) elevations result in breakthroughs in many areas of life.

Classical music – and the masterpieces of its greatest representatives Franz Joseph Haydn, Wolfgang Amadeus Mozart and Ludwig van Beethoven – was born after the emergence of Isaac Newton's classical mechanics and gravity. The phenomena of tonality as gravity in music and the sonata form/ principle as a reflection of dialectics in music appeared as major features of classical music. Newton's classical mechanics also had a profound impact on the development of philosophy, namely John Locke's works, David Hume's empiricism, Immanuel Kant's metaphysics and further concepts that followed them.

The twentieth century has become a turning point in global history. Social upheavals and turmoil, the dissolution of old empires and the birth of new states, groundbreaking scientific discoveries and other events have influenced the deep yet paramount shifts in music, literature and art of the same time period.

Contemporary music/avant-garde (the techniques that revolutionised the tonal system) was born at the turn of the nineteenth and twentieth centuries—right after the exhaustion of the era of romanticism—exactly when fundamental discoveries were made in the world of science, that radically changed our perception of the structure of the universe:

- Splitting the Atom by Ernest Rutherford.

R.-B. Abdyssagin, *Quantum Mechanics and Avant-Garde Music*, https://doi.org/10.1007/978-3-031-63161-0_1

- Albert Einstein's Theory of Relativity and Theory of Gravity.
- Quantum Mechanics by Max Planck, Niels Bohr, Louis de Broglie, Werner Heisenberg, Max Born, Pascual Jordan, Erwin Schrödinger, Paul Dirac, Wolfgang Pauli, Enrico Fermi, et al. (more details in Chapter 2)

This was reflected in music by the emergence of new techniques of composition (Parts II and III of the book) and extended methods of playing instruments (Chapters 20 and 21).

It is impossible to draw an equals sign between music and physics. Nevertheless, certain empirical correlations can be found. Science and art are the highest manifestations of the mind. As Ludwig van Beethoven once said, 'only art and science can raise men to the level of gods'.

None of the reviewed connections are intended to directly relate to the acoustical phenomena of music or to the physical nature of sound. This book researches poetic, figurative, imaginary, metaphoric and artistic correlations between the structure of musical works and certain elements of quantum physics (Part II).

Understanding that sounds are elastic waves is essential and remains the very first point of interconnection between music and physics (Abdyssagin 2013). Like photons are quantised light waves, phonons are quantised sound waves: hence, there is another correlation with quantum physics that touches the true fundamentals of sound and music.

Considering metaphorical correlations, as is known, a map is not the territory itself but only a description of the territory (Covey 2013). Equations and formulas are not nature itself but rather a description of nature. Similarly, a score is not the music itself but only a description of the music. Therefore, a score for music has a similar role to equations and mathematical formalism for physics (more about this can be found in Chapter 12).

Just as every theory in physics has to be tested by experiment in order to be confirmed or refuted, every musical score/compositional theory has to be tested by musical experiment – performance – in order to become music. As well-known, the role of experiment in classical mechanics and quantum mechanics varies, as it does in classical music and in certain techniques of avant-garde music (this is discussed in detail in Chapter 10).

For example, in classical mechanics an experiment does not heavily affect the result, while in quantum mechanics the experiment creates reality. Classical mechanics is deterministic ('Laplace's demon' is a clear example of it), just like the scores of classical music, in the sense that in scores (within the traditional/conventional notation system) of classical music, there is only a single direction, the only possible trajectory of movement from point A to point B. Between any two (imaginary) points in the score only one line can be drawn.

Performance of a classical piece (or better say any work written using classical notation) is always predictable in the sense of the form; while a performer can greatly vary all other parameters of music including tempo, dynamics, articulation, etc. (which partly accounts for differences in interpretation), a performer will never be able to change the arrow of direction, the trajectory of movement in music. And even if the listener is not familiar with the music being performed, looking in the score he/she can always say what was before and predict what will follow, and at one glance he/she has the full form and structure of the work just from the score.

However, that is absolutely not the case with the aleatoric technique: Karlheinz Stockhausen's *Klavierstück XI* (1956) has a score that consists of 19 fragments scattered across a single large page, and a performer has to choose the sequence, the order in which to play them. Thus, a performer (a performance) will create/shape the reality of this music, just as an experiment creates/shapes reality in quantum mechanics. On a global level, the score of this Stockhausen's particular piece is indeterministic, and there is no such thing as a pre-defined form (as in traditional scores), the form of *Klavierstück XI* exists in a superposition of possibilities and probabilities. There is no single pre-defined arrow/direction/trajectory of form and dramaturgy.

In classical notation, there is a pre-defined trajectory independent to performance, while in this piano piece by Stockhausen when there is no performance, there is no single trajectory. When there is no performance/ measurement, we should assume that there are simultaneously a multitude of possible/potential trajectories, just like a quantum object in quantum mechanics (when not measured) moves in all possible directions simultaneously (as beautifully demonstrated in the famous double-slit experiment). Moreover, every performance will render only one of the possible forms of this work, and even if the listener has the score, the score itself (in its inherent ontological global indeterminism of form) will not predict the sequence of events/a trajectory of movement. As analysed in Chapter 13 in greater detail, Stockhausen's *Klavierstück XI* metaphorically exhibits the uncertainty principle.

In my book *Mathematics and Contemporary Music* (2013) I discovered the metaphorical correlation that silence in music is analogous to 0 in mathematics and void in the universe: Silence—0—Void. Silence is the most absolute and impeccable harmony.

And now we come to the idea of the shadows of the void (which is also a title of my music work analysed in Chapter 16). There are many phenomena that are beyond our perception and understanding, which we tend to call the void. But is it really a void? These phenomena (which we call the void) can affect our world in an extremely profound way. These 'void(s)' leave traces in the fabric of reality (just like anything material does), and I call them 'the shadows of the void'. The so-called Dark Energy and Dark Matter are good examples. As is commonly stated now, Dark Energy and Dark Matter together constitute 95% of the total mass-energy in the present-day observable universe! And we do not even know precisely what they are, these mysterious and enigmatic entities. As the NASA website indicates: 'By fitting a theoretical model of the composition of the universe to the combined set of cosmological observations, scientists have come up with the composition [...] ~ 68% dark energy, ~ 27% dark matter, ~ 5% normal matter'.

Therefore, as elucidation of my 'shadows of the void' theory, I propose the following postulates:

- The Void is not nothing; it is less than nothing but more than everything.
- The Void, or 'what does not exist', is more important than what exists, because 'what does not exist' allows the existence of what exists.

Another important aspect is the nature of mathematics. In my view mathematics is the language in which all sciences speak (2013). So the following postulate emerges:

- Mathematics is the first and to-date the most perfect virtual reality, which has been created by humanity in order to describe objective reality.

Music has an impact on humans like nothing else. We feel music with every fibre of our being; it penetrates into the deepest and most hidden corners of the universe within us, human consciousness, where nothing else has access (Abdyssagin 2017, p. 5).

This is also a personal book which reflects my professional activities as a composer and pianist and personal experience as a musician. For example, my compositions *Petrogliff* for two pianos (2011), *Time Run* for symphony orchestra (2018), *A drop of Eternity*, concerto for violin and symphony orchestra (2018), *Tears of Silence*, concerto for piano and symphony orchestra (2018), *The Space of Resonance* for 7 performers, *Ombre del Vuoto* (Shadows of the Void) for 11 performers (2019), *Serenade of Invisible Stars* for 5 performers (2019), *13 Notes from the Parallel Universe* for string quartet (2020), *Quantum Reality* for 12 performers (2020), Symphony №2 *Chaos and Order* for full orchestra + 'audio piano' (2023), etc., are influenced by physics/sciences and explore the phenomena of time; space; resonance; silence etc. Some of the above-mentioned compositions as well as my other music works are analysed in the book as additional illustrations of metaphoric correlations between quantum mechanics and contemporary music.

Technical notes: in Scientific Pitch Notation, C_4 indicates Middle C (Fig. 1.1). According to the current international concert pitch standard, C_4 = 261.63 Hz when A_4 = 440 Hz.

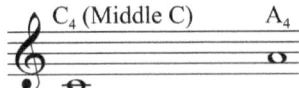

Fig. 1.1 C_4 and A_4

Throughout the book the letter notation (Fig. 1.2) is used according to the German system (the only difference is that sometimes B♭ is used for convenience):

Fig. 1.2 German system of letter notation

The publication of this book is dedicated to symbolic anniversaries that will take place in 2025: the 100th anniversary of quantum mechanics, the 110th anniversary of Einstein's general relativity, the 120th anniversary of Einstein's special relativity, the 125th anniversary of 'energy quanta' and Planck's constant, and the 90th anniversary of Einstein–Podolsky–Rosen (EPR) paradox.

References

Abdyssagin, R.-B. 2013. *Mathematics and Contemporary Music*. Almaty: Kazak Universiteti.

Abdyssagin, R.-B. 2017. *Bozhestvenniy put'* (The Divine Path). Almaty: Kazak Universiteti.

Covey, S. 2013. *The 7 Habits of Highly Effective People*. 25th Anniversary Edition. RosettaBooks LLC.

NASA, n.d. *Dark Energy, Dark Matter*. https://science.nasa.gov/astrophysics/focus-areas/what-is-dark-energy/ (Accessed 19 January 2024).

Stockhausen, K. 1956. *Klavierstück XI*. Vienna: Universal Edition.

2

Correlations in Timeline

A metaphorical correlation between the development of physics and the emergence of new music can be witnessed even from a chronological perspective (Abdyssagin 2022, pp. 8–9). Here events are matched with one other, juxtaposing the timeline of discoveries. This does not imply any direct connection between specific events. However, the fact that breakthroughs in physics and in music happened nearly at the same time seems to be no coincidence. It was impossible for contemporary music not to emerge. Therefore, the birth of new music was not random or accidental, but truly an inevitable event. As Vlatko Vedral pointed out: 'notions of beauty and truth in one area inevitably affect ideas in another' (2018, p. 7).

In 1897 the British physicist Joseph John Thomson (Fig. 2.1) presented his discovery of electron to the world, thus revealing that the atom has a structure (electron was the first subatomic particle to be discovered). In 1899 the Austrian composer and founder of the Second Viennese School Arnold Schoenberg (Fig. 2.2) created *Verklärte Nacht* (Transfigured Night) Op. 4—an extraordinary piece of music still in a romantic style with an atmosphere of expressionism, which is, nevertheless, distinguished by a deliberately liberal attitude towards traditional tonality (Fig. 2.3).

It is interesting to mention that Joseph John Thomson received the 1906 Nobel Prize in Physics for the discovery that electron is a particle, while his son George Paget Thomson (also a physicist) received the 1937 Nobel Prize in Physics for the discovery that electron is a wave. This beautifully shows the wave-particle duality, one of the most important concepts in quantum

© The Author(s), under exclusive license to Springer Nature
Switzerland AG 2024
R.-B. Abdyssagin, *Quantum Mechanics and Avant-Garde Music*,
https://doi.org/10.1007/978-3-031-63161-0_2

Fig. 2.1 J.J. Thomson (*Source* Wikimedia Commons)

Fig. 2.2 A. Schoenberg (Photo by Man Ray. *Source* Wikimedia Commons)

mechanics. It states that (from the point of view of classical physics) an electron sometimes behaves like (has the properties of) a wave, and sometimes behaves like (has the properties of) a particle. While in fact, it is 'neither and both' (Davies 1989, xiv). Paraphrasing Werner Heisenberg (2000), it can be said that electron is both wave and particle, and neither wave nor particle. The answer is that it is a micro-object with peculiar properties.

In 1900 the German theoretical physicist Max Planck (Fig. 2.4) discovered energy quanta (while working on black-body radiation) and formulated Planck's law, which symbolised the birth of modern physics. He also introduced Planck's constant (h), which is fundamentally important for quantum mechanics. Planck's constant is prominently present in Planck's energy-frequency relation:

$$E = h\upsilon$$

In this foundational equation E stands for the energy, υ for the frequency, while h is Planck's constant. In 1902 the Dutch physicists Hendrik Antoon Lorentz and Pieter Zeeman shared the Nobel Prize in Physics. They are famous for the Lorentz transformations (1899) and the Zeeman effect (1896) respectively. In 1903–1905 the French impressionist and symbolist composer Claude Debussy (Fig. 2.7) created *La Mer, trois esquisses symphoniques pour orchestre* which can be perceived as one of the first works that uses new techniques of orchestration and implements timbral dramaturgy and introduces the timbral-textural dimension of harmony.

Fig. 2.3 Excerpt from the manuscript of A. Schoenberg *Verklärte Nacht* Op. 4 (Public domain. *Source* IMSLP, Petrucci Music Library)

Fig. 2.4 M. Planck (*Source* Wikimedia Commons)

Fig. 2.5 H.A. Lorentz (*Source* Wikimedia Commons)

Fig. 2.6 25 year old A. Einstein (Photo by Lucien Chavan. *Source* Wikimedia Commons)

1905 was Albert Einstein's *Annus mirabilis* (miracle year), when—at the age of 26—he published 4 groundbreaking papers in *Annalen der Physik* journal. In one of these articles Einstein introduced the mass-energy equivalence, which became the most famous equation in the world:

$$E = mc^2$$

These *Annus mirabilis* articles of Einstein (Fig. 2.6) revolutionised physics and changed our perception of fundamental concepts such as space, time, energy and mass.

In 1908 Charles Ives (Fig. 2.8) composed the iconic work *The Unanswered Question* (later revised in 1930–1935)—a revolutionary composition in the history of American contemporary music. In the same 1908 the French composer Maurice Ravel (Fig. 2.9) wrote *Gaspard de la nuit* which channeled some newly born concepts of piano expressivity and augmented the relative fundamentals of tonality. A. Schoenberg composed *Drei Klavierstücke* Op. 11 in 1909—a cycle of innovative pieces in which utter expressionism and usage of a peculiar piano technique declared the eclipse of romantic architectonics and predicted the inception of new music.

In 1911 the physicist Ernest Rutherford (Fig. 2.10) published the article *The Scattering of α and β Particles by Matter and the Structure of the Atom* where the concept of 'atomic nucleus' was first introduced, in the journal *Philosophical Magazine*. In 1912 A. Schoenberg created *Pierrot lunaire* Op. 21 where the Sprechstimme/Sprechgesang technique was applied in addition to the new concepts of instrumental-vocal relations.

In 1913 the Danish physicist Niels Bohr (Fig. 2.11) presented his model of hydrogen atom with a small nucleus in the centre and electrons moving around in the fixed orbits (resembling the structure of the Solar System). Bohr's work became an immediate success. Even Einstein was astonished at Bohr's achievement, and said that it is 'the highest form of musicality in the sphere of thought' (Wilczek 2021, p. 108). Bohr's model was later refined by Arnold Sommerfeld. Although Bohr's orbital model was later replaced, at that time it was a fascinating breakthrough that advanced the early stages of quantum mechanics.

In the same year 1913 Igor Stravinsky's (Fig. 2.12) epochal ballet *Le Sacre du printemps* (The Rite of Spring) pas premiered in Paris; this multidimensional many-layered composition was a breakthrough and revolutionary revelation of the possibilities of new music and even new art in general. The premiere of Stravinsky's ballet was one of the turning points and most significant events in the history of twentieth-century art. In fact, evaluating metaphorically, the premiere of *Le Sacre du printemps* opened the

Fig. 2.7 C. Debussy (Photo by Nadar. *Source* Wikimedia Commons)

Fig. 2.8 C. Ives (*Source* Wikimedia Commons)

Fig. 2.9 M. Ravel (*Source* Gallica—Bibliotheque nationale de France)

Fig. 2.10 E. Rutherford (*Source* Library of Congress. George Grantham Bain Collection)

Fig. 2.11 N. Bohr (*Source* Library of Congress. George Grantham Bain Collection)

Fig. 2.12 I. Stravinsky (*Source* Library of Congress. George Grantham Bain Collection)

gates to the then unknown and yet-to-be-discovered universe of new music. Even nowadays it is hard to overestimate or exaggerate the influential role of this ballet in the development of modern art.

In 1914 James Franck and Gustav Ludwig Hertz performed an experiment (later called the Franck-Hertz experiment) which was direct experimental proof of the quantum nature of the atom. 'Our understanding of the world was transformed by the results of this experiment; it is arguably one of the most important foundations of the experimental verification of the quantum nature of matter. The significance of this work was recognised by the award to James Franck and Gustav Hertz of the 1925 Nobel Prize in Physics' (Rice and Jortner 2010, pp. 6–7).

In 1915 the General theory of relativity (also referred to as Einstein's theory of gravity) was published by A. Einstein. Between 1911 and 1913 Alexander Scriabin composed his 5 last piano sonatas: No. 6, No. 7 *White Mass*, No. 8, No. 9 *Black Mass* and No. 10 (Fig. 2.13). These sonatas are characterised by an innovative approach to structuring form and tonality. In early works of Scriabin tonality served more or less like in any other late-romantic music, whereas in his last piano sonatas tonality itself transforms into a new kind of 'gravity' that does not directly interfere with dramaturgy, but indirectly influences the global form, distantly and imaginatively resembling the space–time curvature in Einstein's General relativity theory.

A. Schoenberg composed the *Suite for Piano* Op. 25 in 1921/23 which appeared to be the first absolute implementation of the twelve-tone technique introducing substantial novelty and the silent revolution to music. As one of the greatest composers and conductors of the twentieth century Pierre Boulez wrote: 'with the 12-tone system, music moved out of the world of Newton and into the world of Einstien' (May 2023, p. 67).

In 1924 the French physicist Louis de Broglie (Fig. 2.14) in his PhD thesis 'showed that a certain matter wave could 'correspond' to a moving electron, just as a light wave corresponds to a moving light quantum' (Heisenberg 2000, pp. 8–9). For the first time the expression 'Quantum Mechanics' appeared in Max Born's (Fig. 2.15) article *Über Quantenmechanik* published in *Zeitschrift für Physik* in 1924.

In 1924 Satyendra Nath Bose's article *Plancks Gesetz und Lichtquantenhypothese* was published in *Zeitschrift für Physik* under German translation by A. Einstein, and Einstein also published his own article in support of Bose's ideas in the same issue of the journal. This was the introduction of Bose–Einstein statistics (for bosons). Its counterpart, Fermi–Dirac statistics (for fermions) was independently introduced in 1926 in Enrico Fermi's article *Sulla quantizzazione del gas perfetto monoatomico* and Paul Dirac's article *On the Theory of Quantum Mechanics*.

In 1925 Werner Heisenberg published his breakthrough article *Über quantentheoretische Umdeutung kinematischer und mechanischer Beziehungen* (Quantum theoretical re-interpretation of kinematic and mechanical relations) in *Zeitschrift für Physik*. This laid the groundwork for the Matrix

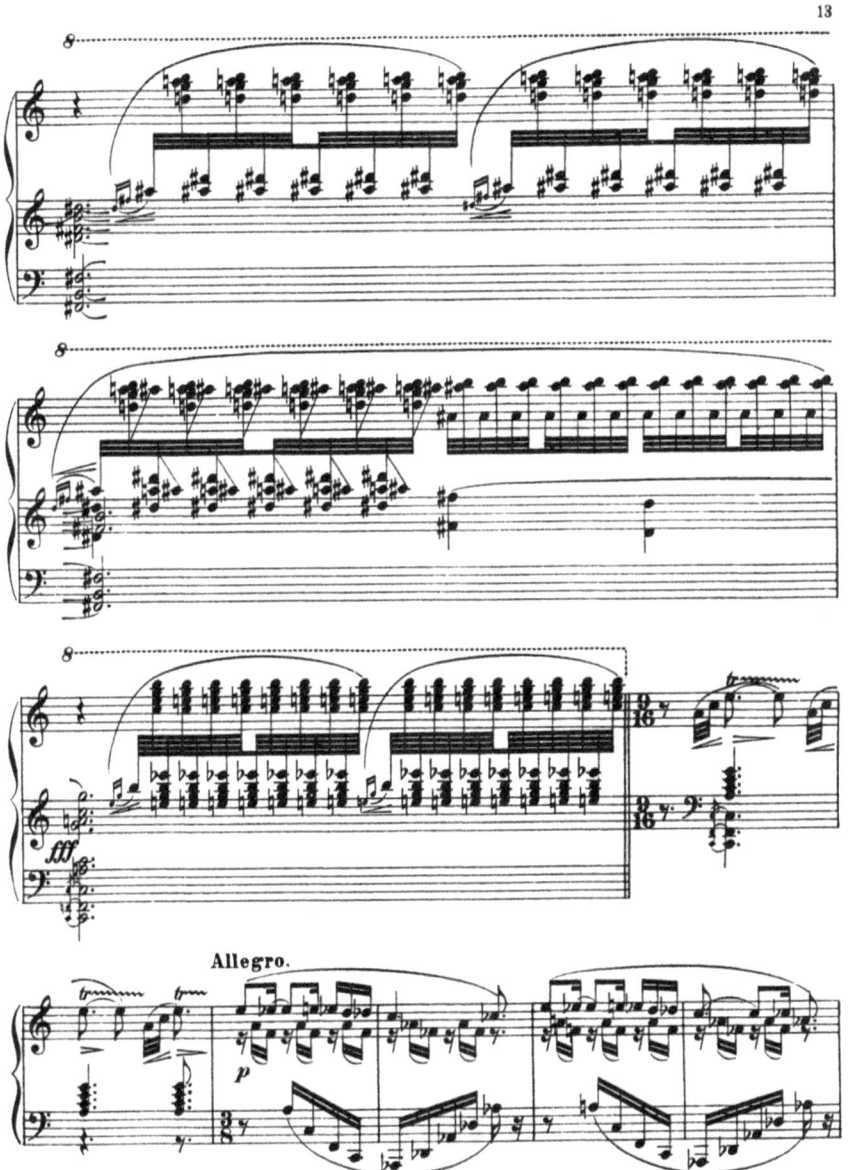

Fig. 2.13 Excerpt from the score of A. Scriabin Piano Sonata No. 10, Op. 70 (Moscow: P. Jurgenson. Public domain. *Source* IMSLP, Petrucci Music Library)

Fig. 2.14 L. de Broglie (*Source* Nobel Foundation)

Fig. 2.15 M. Born (*Source* Wikimedia Commons)

mechanics formulation of quantum mechanics. The development of matrix mechanics representation of quantum mechanics was then immediately followed by Max Born and Pascual Jordan. In the same 1925 year the famous 'Pauli exclusion principle' was first formulated for electrons (later extended to all fermions in 1940) by the 25-year-old Austrian theoretical physicist Wolfgang Pauli.

In 1925 Edgard Varèse published the composition *Integrales* for wind, brass and percussion, and created the symphonic poem *Arcana* in 1925–1927 (later revised in 1931–1932) where totally new approaches to composing music were used together with an unprecedented hierarchy of music parameters led by the sound clusters.

In 1927 the German physicist Werner Heisenberg formulated his famous uncertainty principle in the article *Über den anschaulichen Inhalt der quantentheoretischen Kinematik und Mechanik* published in *Zeitschrift für Physik*. Notably, Heisenberg was only 25 years old at that time (Fig. 2.16).

In 1925 Erwin Schrödinger (Fig. 2.17) formulated the famous Schrödinger equation, and published it in 1926, thus presenting the wave mechanics formulation of quantum mechanics:

$$i\hbar\frac{\partial}{\partial t}\Psi = \hat{H}\Psi$$

In 1928 Paul Dirac (Fig. 2.18) (at the age of only 26 years old) derived the relativistic wave equation which later became known as the Dirac Equation.

Fig. 2.16 W. Pauli, W. Heisenberg and E. Fermi on a boat on Lake Como, Italy, September 1927 (Photo: CERN/© Franco Rasetti. *Source* Nobel Foundation)

Fig. 2.17 E. Schrödinger (*Source* Nobel Foundation)

Fig. 2.18 P.A.M. Dirac (*Source* Nobel Foundation)

These chronological interconnections also continued into the 1930s. Dirac published his mathematical formalism of quantum mechanics (with operators on a Hilbert space) in 1930 in a book called *The Principles of Quantum Mechanics*. In 1932 John von Neumann published *Mathematische Grundlagen der Quantenmechanik* (Mathematical Foundations of Quantum Mechanics). Together these publications form the Dirac–von Neumann axioms and establish the mathematical formalism of quantum mechanics. Later Dirac

Fig. 2.19   K. Gödel (*Source* Wikimedia Commons)

Fig. 2.20   A. Webern (Photo by Georg Fayer. *Source* Österreichische Nationalbibliothek)

developed Bra–ket notation in his publication *A New Notation for Quantum Mechanics* (1939).

In 1931 in article called *Über formal unentscheidbare Sätze der Principia Mathematica und verwandter Systeme I* (On Formally Undecidable Propositions of Principia Mathematica and Related Systems I) a 25-year-old mathematician and logician Kurt Gödel (Fig. 2.19) published his famous Gödel's incompleteness theorems which revolutionised mathematics and logic. It 'contained one of those landmark intellectual achievements, like Heisenberg's uncertainty principle or the discovery of the structure of DNA, that fundamentally change people's view of things' (Moore 2022, p. xix). The initial informal formulation of Gödel's theorem states that 'no axiomatisation can determine the whole truth and nothing but the truth concerning arithmetic' (Moore 2022, p. 3). It is worth noting that Gödel's second theorem's generalised version can be stated as 'no consistent, sufficiently strong, axiomatisable theory can contain a statement corresponding to a statement of its own consistency' (Moore 2022, p. 93).

One of the most captivating facets of Gödel's theorem was noted by physicist Tim Palmer: 'The key result at the heart of Gödel's theorem is a formula which can be constructed from the rules of any sufficiently powerful logical system. However, it is a strange, self-referential formula. It states that if this formula could be proved, the formula would be false' (2022, p. 45). Reference to self-referentiality is also exceedingly relevant when drawing connections with music, because many aspects of different musical

Fig. 2.21 First bars of A. Webern *Variations Op. 27*, published by Universal Edition (© By kind permission of Universal Edition A.G., Wien)

systems—from polyphony, harmony and tonality to the twelve-tone technique and integral serialism and other more advanced directions—are utterly self-referential. The meaning and role of entities within such systems are also commonly determined by the self-referential power of the system.

In 1935 Albert Einstein, Boris Podolsky and Nathan Rosen published a paper *Can Quantum–Mechanical Description of Physical Reality be Considered Complete?* where they introduce a thought experiment known as the Einstein–Podolsky–Rosen (EPR) paradox. EPR as well as Schrödinger's subsequent works (1935a, 1935b, 1936) gave rise to quantum entanglement which is one of the milestones of quantum information science and quantum computing (more on this in Chapter 21).

In 1936 the Austrian composer Anton Webern (Fig. 2.20) wrote his fundamental oeuvre Variations for piano Op. 27 (Fig. 2.21), thus welcoming the serialism and structuralism which will later lead the mainstream of twentieth century music. Metaphorically Webern's *Variations* Op. 27 played a similar role in music to that of Gödel's theorem in mathematics. It can be argued that Webern showed the 'incompleteness' of Schoenberg's early dodecaphonic approach and invented his own system reaching far beyond what was deemed possible and feasible in the scope of established dodecaphony.

However, as philosopher Adrian W. Moore pointed out (during our correspondence in January 2024), that would suggest that 'Gödel showed the incompleteness of one particular system, whereas he showed something far stronger than that: he showed that completeness of a certain kind in arithmetic was unobtainable'. Therefore, as Professor Moore states, to have an appropriate stance to draw a genuine analogy between Webern's Variations and Gödel's theorem, it would need to be argued 'that Webern's approach was incomplete too, and indeed that any possible approach of this kind to music would always be incomplete'. In theory, conjecturing and hypothesizing in that direction could be possible, nevertheless, whether it will be true or not (and whether any technical/syntactic/semantic/ideological proof can be found) remains a question, and this question is beyond the scope of this chapter.

Webern's twelve-tone structure developed more complicated and sophisticated interconnections between various thematic elements and nonlinear correlations between different structural parameters of music. While Schoenberg (even in his dodecaphonic works) relied on Beethoven's formal language and Beethovenian structures, Webern went further and through the prism of medieval polyphonic systems developed new structural dimensions within the twelve-tone system. This allowed the emergence of significant works such as Oliver Messiaen's *Quatre Études de rythme* (also known as 'Rhythm Etudes') and Pierre Boulez's Second Piano Sonata which manifested integral (total) serialism. Stretching this figurative connection, it can be imagined that Webern in a certain sense even displayed nonlocality in music, when systems in diverse and distant parts of the work remain correlated (or even entangled) and connected through tone rows and serial structures, as well as combinatorial functions of rhythms and dynamics. While certain aspects of nonlocal correlations are present in many musical works throughout history, Webern's systematic and structural approach elevates them to a completely new height and profound depth that determine musical architectonics.

Similar processes also occurred in the visual arts and literature. For example, in 1915—the same year that A. Einstein published his general theory of relativity—Kazimir Malevich painted his famous *Black Square* (Fig. 2.22). A black square painted on white even now remains one of the greatest and highest exaltations of abstract art, a supreme manifestation of modern art in general. This painting became the symbol of suprematism and the whole history of contemporary visual art can be separated into 'before' and 'after' the *Black Square*.

The *Black Square* itself is an idealisation, a picture of an absolute and all-penetrating void, a kind of a black hole, a metaphorical visualisation of gravitational singularity, or even a picture of true zero. As K. Malevich himself said: 'It is from zero, in zero, that the true movement of being begins' (Marcadé 2003, p. 40). Malevich also intended to establish a new theory of nonobjective art and suprematism in order to 'go beyond zero' (Malevich 1976, p. 186, cited in Gurianova 2003, pp. 45–46). The movement

Fig. 2.22 K. Malevich *Black Square* (Tretyakov Gallery, Moscow. *Source* Wikimedia Commons)

of suprematism also touched modern music, and involved such figures as the composer Nikolai Roslavets and many others (Gurianova 2003, pp. 51, 55).

Considering literature, James Joyce published his *Ulysses* in 1922 in Paris. This novel became one of the pillars of modernist literature, an eternal symbol and embodiment of the 'stream of consciousness' technique, a personification of a new age of literature. Declan Kiberd called *Ulysses* 'modernism's most sociable masterpiece' (2009).

This novel was highly influential not only in literary arts, but also in music: one of the most important twentieth century creators, the Italian composer Luciano Berio composed *Thema (Omaggio a Joyce)*—'electro-acoustic elaboration of Cathy Berberian's voice on tape' (1958). Umberto Eco also contributed to this composition.

In J. Joyce's *Ulysses* the uncertainty and indeterminacy find intrinsic and idiomatic literary incarnation—in the last *Episode 18, Penelope* where no punctuation mark is present at all and the whole text consists of thoughts without interruptions and control, a pure stream of consciousness (1922). This lack of punctuation and holistic indeterminacy can provide allegoric links with the aleatoric works of John Cage, Morton Feldman and others.

In general, the 1920s were remarkable years in the literature. From 1913 until 1927 French author Marcel Proust published his 7-volume novel *À la recherche du temps perdu* (In Search of Lost Time) which became one of the paramount and most influential works of the entire twentieth century. In 1920 Agatha Christie published her first detective novel—*The Mysterious Affair at Styles*—where she introduced her iconic character Hercule Poirot. In 1922 Thomas Stearns Eliot (who was awarded the 1948 Nobel Prize in Literature) published a poem *The Waste Land*—which is considered as not only the central piece of modernist poetry, but also as one of the major achievements of English language poetry in the twentieth century in general. The Modernist English writer Virginia Woolf published her influential novel *Mrs. Dalloway* in 1925, while in the same year in America, Francis Scott Fitzgerald's legendary *The Great Gatsby* was published. 1926 is marked by the publication of the first novel by Ernest Hemingway, winner of the 1954 Nobel Prize in Literature, *The Sun Also Rises* (also known as *Fiesta*), and Thomas Wolfe published his first novel *Look Homeward, Angel* in 1929. The British author Arthur Conan Doyle wrote about Sherlock Holmes—probably the most famous fictional character ever—from 1887 until 1927. The list of similar important developments can be continued.

The 1920s were also fruitful years for philosophy—notable examples include Ludwig Wittgenstein's *Logisch-Philosophische Abhandlung* (Tractatus Logico-Philosophicus), published in 1921, one of the foremost works of the tradition of analytic philosophy, and Martin Heidegger's *Sein und Zeit* (Being and Time), published in 1927, a great monument of continental philosophy and existentialism. In fact, during that time Martin Heidegger 'was aware of developments in modern physics' (Carson 2014, p. 89). Werner Heisenberg was also aware of developments in analytic philosophy and was interested in the philosophy of language, and—recalling occasional encounters with Bertrand Russell (and discussion about Ludwig Wittgenstein's *Tractatus Logico-Philosophicus*)—wrote that 'Russell loved the Tractatus and told me that he could not understand what Wittgenstein meant in his later work. I told him that I felt the Tractatus was wrong or trivial in all essential points, but I had a high esteem of the late Wittgenstein' (Heisenberg 1974, cited in Carson 2014, p. 88).

In general, it was a time when all major areas of human thought and activity were revolutionised. A time of breakthroughs in the worldview and the consciousness of the humanity, when revolutionary ideas penetrated and captured the minds of brilliant people across many disciplines and directions. As mathematician Richard Jozsa said (during our meeting on 6 October 2023): 'It was the force of time: the time itself influenced the great minds to make discoveries. It was a time when breakthroughs and revolutionary discoveries happened not only in physics and sciences, but also in almost all areas, including fine arts and music'.

References

Abdyssagin, R.-B. 2022. *Noveishie kompozitorskie i ispolnitel'skie tehniki* (The Newest Composing and Performing Techniques). Astana: KazNUA.

Berio, L. 1958. *Thema (Omaggio a Joyce)—Author's note*. Centro Studi Luciano Berio. http://www.lucianoberio.org/thema-omaggio-a-joyce-authors-note?948 448529=1 (Accessed 12 March 2023).

Born, M. 1924. Über Quantenmechanik. *Zeitschrift Für Physik* 26: 379–395.

Bose, S.N. 1924. Plancks Gesetz und Lichtquantenhypothese. *Zeitschrift Für Physik* 26 (1): 178–181.

Boulez, P. 1950. *Piano Sonata No. 2*. Paris: Heugel et Cie.

Carson, C. 2014. *Heisenberg in the Atomic Age: Science and the Public Sphere*. New York: Cambridge University Press.

Christie, A. 1920. *The Mysterious Affair at Styles*. New York: John Lane.

Davies, P. 1989. Introduction. In *Physics and Philosophy*, W. Heisenberg, vii–xvii. Penguin Classics.

de Broglie, L. 1924. *Recherches sur la théorie des quanta*. PhD Thesis, Paris.

Debussy, C. 1905. *La Mer, trois esquisses symphoniques pour orchestre* (The Sea, Three Symphonic Sketches for Orchestra), L. 109, CD. 111. Paris: Éditions Durand.

Dirac, P.A.M. 1926. On the Theory of Quantum Mechanics. *Proceedings of the Royal Society A* 112 (762): 661–677.

Dirac, P.A.M. 1930. *The Principles of Quantum Mechanics*. Oxford: Clarendon Press.

Dirac, P.A.M. 1939. A New Notation for Quantum Mechanics. *Mathematical Proceedings of the Cambridge Philosophical Society* 35 (3): 416–418.

Drutt, M., ed. 2003. *Kazimir Malevich: Suprematism*. New York: The Solomon R. Guggenheim Foundation.

Einstein, A. 1905. Zur Elektrodynamik bewegter Körper. *Annalen der Physik* 17 (10): 891–921.

Einstein, A., B. Podolsky, and N. Rosen. 1935. Can Quantum-Mechanical Description of Physical Reality Be Considered Complete? *Physical Review* 47 (10): 777–780.

Eliot, T.S. 1922. *The Waste Land*. New York: Boni & Liveright.

Fermi, E. 1926. Sulla quantizzazione del gas perfetto monoatomico. *Rendiconti Lincei* 3: 145–149.

Fitzgerald, F.S. 1925. *The Great Gatsby*. New York: Scribner.

Franck, J., and G. Hertz. 1914. Über Zusammenstöße zwischen Elektronen und Molekülen des Quecksilberdampfes und die Ionisierungsspannung desselben. *Verhandlungen der Deutschen Physikalischen Gesellschaft* 16: 457–467.

Gödel, K. 1931. Über formal unentscheidbare Sätze der Principia Mathematica und verwandter Systeme I. *Monatshefte Für Mathematik und Physik* 38: 173–198.

Gurianova, N. 2003. The *Supremus* "'Laboratory House'": Reconstructing the Journal. In *Kazimir Malevich: Suprematism*, ed. M. Drutt, 45–59. New York: The Solomon R. Guggenheim Foundation.

Heidegger, M. 1927. *Sein und Zeit*. Tübingen: Max Niemeyer Verlag.

Heisenberg, W. 1925. Über quantentheoretische Umdeutung kinematischer und mechanischer Beziehungen. *Zeitschrift Für Physik* 33 (1): 879–893.

Heisenberg, W. 1927. Über den anschaulichen Inhalt der quantentheoretischen Kinematik und Mechanik. *Zeitschrift Für Physik* 43 (3–4): 172–198.

Heisenberg, W. 2000. *Physics and Philosophy*. Penguin Classics.

Hemingway, E. 1926. *The Sun Also Rises*. New York: Scribner.

Ives, C. 1953. *The Unanswered Question*. New York: Southern Music Publishing Co., Inc.

Joyce, J. 1922. *Ulysses*. Paris: Shakespeare and Company.

Kiberd, D. 2009. *Ulysses, Modernism's Most Sociable Masterpiece*. London: The Guardian. https://www.theguardian.com/books/2009/jun/16/

Marcadé, J.C. 2003. Malevich, Painting, and Writing: On the Development of a Suprematist Philosophy. In *Kazimir Malevich: Suprematism*, ed. M. Drutt, 33–43. New York: The Solomon R. Guggenheim Foundation.

May, A. 2023. *The Science of Music: How Technology Has Shaped the Evolution of an Artform*. London: Icon Books Ltd., Hot Science series.

Messiaen, O. 1950. *Quatre Études de rythme*. Paris: Éditions Durand.

Moore, A.W. 2022. *Gödel's Theorem: A Very Short Introduction*. Oxford: Oxford University Press.

Palmer, T. 2022. *The Primacy of Doubt*. Oxford: Oxford University Press.

Pauli, W. 1925. Über den Zusammenhang des Abschlusses der Elektronengruppen im Atom mit der Komplexstruktur der Spektren. *Zeitschrift Für Physik* 31 (1): 765–783.

Planck, M. 1900a. Über eine Verbesserung der Wienschen Spektralgleichung. *Verhandlungen der Deutschen Physikalischen Gesellschaft* 2: 202–204.

Planck, M. 1900b. Zur Theorie des Gesetzes der Energieverteilung im Normalspectrum. *Verhandlungen der Deutschen Physikalischen Gesellschaft* 2: 237.

Planck, M. 1900c. Entropie und Temperatur strahlender Wärme. *Annalen der Physik*. 306 (4): 719–737.

Planck, M. 1900d. Über irreversible Strahlungsvorgänge. *Annalen der Physik* 306 (1): 69–122.

Ravel, M. 1908. *Gaspard de la nuit* (Trois poèmes pour piano d'après Aloysius Bertrand) M. 55.

Rice, S.A., and J. Jortner. 2010. *James Franck 1882–1964: A Biographical Memoir*. Washington, DC: National Academy of Sciences.

Rutherford, E. 1911. 'The Scattering of α and β Particles by Matter and the Structure of the Atom'. *The London, Edinburgh, and Dublin Philosophical Magazine and Journal of Science*, 21 (125): 669–688. Taylor & Francis.

Schoenberg, A. 1899. *Verklärte Nacht*, Op. 4, Sextett für zwei Violinen, zwei Violen und zwei Violoncelli. Berlin: Verlag Dreililien, Richard Birnbach.

Schoenberg, A. 1910. *Drei Klavierstücke*, Op. 11. Vienna: Universal Edition.

Schoenberg, A. 1914. *Pierrot lunaire*, Op. 21. Vienna: Universal Edition.

Schoenberg, A. 1925. *Suite for Piano*, Op. 25. Vienna: Universal Edition.

Schrödinger, E. 1926. Quantisierung als Eigenwertproblem. *Annalen der Physik* 384 (4): 273–376.

Schrödinger, E. 1935a. Discussion of Probability Relations Between Separated Systems. *Mathematical Proceedings of the Cambridge Philosophical Society.* 31 (4): 555–563.

Schrödinger, E. 1935b. Die gegenwärtige Situation in der Quantenmechanik. *Naturwissenschaften* 23 (48): 807–812.

Schrödinger, E. 1936. Probability Relations Between Separated Systems. *Mathematical Proceedings of the Cambridge Philosophical Society* 32 (3): 446–452.

Scriabin, A. 1912. *Piano Sonata No. 6*, Op. 62. Berlin: Russischer Musikverlag.

Scriabin, A. 1913a. *Piano Sonata No. 8*, Op. 66. Moscow: P. Jurgenson.

Scriabin, A. 1913b. *Piano Sonata No. 9*, Op. 68, *Messe noire* (Black Mass). Moscow: P. Jurgenson.

Scriabin, A. 1913c. *Piano Sonata No. 10*, Op. 70. Moscow: P. Jurgenson.

Scriabin, A. 1931. *Piano Sonata No. 7*, Op. 64, *Messe blanche* (White Mass). Moscow: Muzgiz.

Stravinsky, I. 1921. *Le Sacre du printemps* (The Rite of Spring). Berlin: Russischer Musikverlag G.m.b.H.

Vedral, V. 2018. *Decoding Reality: The Universe as Quantum Information.* Oxford: Oxford University Press.

von Neumann, J. 1932. *Mathematische Grundlagen der Quantenmechanik.* Berlin: Springer Verlag.

Webern, A. 1937. *Variations for Piano*, Op. 27. Vienna: Universal Edition.

Wilczek, F. 2021. *Fundamentals: Ten Keys to Reality.* Penguin Books.

Wittgenstein, L. 1921. *Logisch-Philosophische Abhandlung*, ed. Wilhelm Ostwald. Annalen der Naturphilosophie, 14.

Wolfe, T. 1929. *Look Homeward, Angel: A Story of the Buried Life.* New York: Scribner.

Woolf, V. 1925. *Mrs. Dalloway.* London: Hogarth Press.

3

On Some Analogies

In general, relations between mathematics and music have been a subject of research and discussion for a long time since Ancient Greece (e.g. Pythagoras). As Steven Weinberg stated, 'the phenomenon that was studied earliest using methods of arithmetic may have been music' (2016, p. 15). Weinberg even suggested (speaking about Pythagoreans) that 'their emphasis on mathematics may have stemmed from an observation about music' (2016, p. 16). Referring to Pythagoreans, Heisenberg wrote that 'for them the simple mathematical ratio between the length of the strings *created* the harmony in sound' (2000, p. 33). It is widely accepted that Pythagoras found that harmonies pleasant to the ear occur if the length of the sounding strings is like the first four integers: 1:2, 2:3, 3:4 (Abdyssagin 2013, p. 7). The detailed role of mathematics (wave equation and ratios) in harmony has been described by Ian Stewart in his book *17 Equations That Changed the World* (2013, pp. 133–137).

Many great scientists wrote about music, for example Johannes Kepler (Fig. 3.1) – development of ancient idea of *Musica universalis* in *Harmonices Mundi*; René Descartes (Fig. 3.2)—*Musicæ Compendium*, which he wrote at the age of 21; Marin Mersenne—*Harmonie universelle* (1636) and Mersenne's laws that describe frequencies, oscillations and harmonies of a vibrating string and are used in instrument construction and tuning; Leonhard Euler (Fig. 3.3)—his habilitation work *Dissertatio physica de sono* (1727), written at the age of 19, then *Tentamen novae theoriae musicae* (1739) and *Tonnetz* theory, etc. Even Isaac Newton himself was interested in analogies with music (Pesic 2014, pp. 121–131). Interestingly, L. Euler first introduced his mathematical binary logarithms in music theory as a way of subdividing

R.-B. Abdyssagin, *Quantum Mechanics and Avant-Garde Music*, https://doi.org/10.1007/978-3-031-63161-0_3

Fig. 3.1 J. Kepler (by August Köhler. *Source* Wikimedia Commons)

Fig. 3.2 R. Descartes (Portrait after Frans Hals, Louvre Museum. *Source* Wikimedia Commons)

Fig. 3.3 L. Euler (Portrait by Jakob Emanuel Handmann, Kunstmuseum Basel. *Source* Wikimedia Commons)

octaves. Only after the first 'musical implementation' ('premiere') did the binary logarithms appear in mathematics.

One of the most important French composers Jean-Phillippe Rameau said: 'Music is a science which should have definite rules; these rules should be drawn from an evident principle; and this principle cannot really be known to us without the aid of mathematics' (Tsuji and Müller 2022, p. 23).

In contemporary music it is crucial to mention one of the most influential composers of the twentieth century Iannis Xenakis who widely used mathematical methods and the computation of probability in his music and documented this approach in the book *Formalized Music: Thought and Mathematics in Composition* (1992), where he described the methodology, apparatus, conceptual fundament, philosophy of stochastic music and its programmed mechanisms. Additionally, Xenakis meditated on topics such as Boltzmann's and Shannon's entropy (p. 255), phonons, quantum mechanics and quantified space and time (pp. 256–257). The work of Xenakis remains one of the greatest evidences of an artistic understanding of profound connections between the worlds of physics and mathematics and the realm of music. He was also a prominent architect, and his expertise in the field of visual arts influenced his musical personality.

For example, in one of his first major works—*Metastaseis[B]*—Xenakis musically depicts the Einstein's space–time curvature. changes in musical density, register, texture and sonoristic layers affect the perception of time and form, metaphorically like matter curving space–time! The matter in the form of sonoristic mass not only 'curves' musical space–time, but also 'replaces' the conventional incarnations of melody/harmony and even rhythm by the force of its sonoristic 'gravity'. This is, in fact, a well-known approach for analysing this majestic yet unearthy composition.

Further information about influence from physics and architecture in this symphonic opus of Xenakis can be found in Richard Taruskin's *Music in The Late Twentieth Century: The Oxford History of Western Music* (2010, pp. 77–81), as part of chapter dealing with indeterminacy and John Cage (pp. 55–101). He also writes about 'Integrated musical time/space' when considering Milton Babbitt's *Composition for* Four Instruments (pp. 140–148).

Additionally, elucidating differences between European (Darmstadt) and American (Princeton) approach to serialism, Taruskin mentions connections between Babbitt's serialism and mathematical 'set theory' (as well as Babbitt's coining the term 'pitch class') in the same book (pp. 136–137). According to Taruskin, Babbitt was well trained not only in music but also in mathematics and formal logic. He immediately realised the potential of formalising and rationalising the twelve-tone composition technique based on set-theory in mathematics. Babbitt formulated his revolutionary theoretical approach in a paper called "The Function of Set Structure in the Twelve-Tone System" (1946), which he submitted for admission to the degree of Doctor of Philosophy (PhD) at the Music Department of Princeton University. Taruskin reminds amusing details that because then in the department there was neither any person qualified enough to evaluate Babbitt's seminal work nor an officially established PhD programme in music, Babbitt received an endowed chair and prestigious professorship. As Taruskin concludes, Babbitt's 'unaccepted Ph.D. dissertation, circulating widely in typescript, became perhaps the most influential unpublished document in the history of twentieth century music'.

Set theory is one of the fundamental branches of mathematics, initially pioneered by German mathematicians Georg Cantor and Richard Dedekind. Unlike other fields of mathematics which emerged gradually throughout the generations of mathematicians, the set theory was born in Cantor's single paper *On a Property of the Collection of All Real Algebraic Numbers* published in 1874. While Dedekind is famous for *Dedekind Cut*—a way of constructing the real numbers from the rational numbers—Cantor is credited not only for his instrumental role in establishing set theory but also for the fact that he proved that the infinity of the continuum is greater than the infinity of natural numbers. This was called the *cardinality of the continuum*. In fact, it was the set-theory that gave rise to Bertrand Russell's famous paradox and Kurt Gödel's *Incompleteness Theorem*.

Douglas Hofstadter's fundamental book *Gödel, Escher, Bach: an Eternal Golden Braid* (1979) deserves special attention. Although many people think that it is only about the works of Gödel, Escher, and Bach, and/or about connections between mathematics, arts and music—this is far from truth (although a great deal of illuminating correlations are indeed revealed). *Gödel, Escher, Bach* is about the rise of consciousness, the nascence of a mind, the emergence of 'I' and 'self'. It is one of the most deep and profound treatises ever written on the phenomenon of self-reference, which describes how seemingly meaningless symbols and entities gain meaning, 'life' and soul (*anima*) within a system constructed from such entities, and this structure/form and interrelations between entities/symbols saturate these symbols/entities with life and meaning. This book reveals the insights into the birth of cognition and even meditates on how neurons in the brain create the mind. Additionally, through thorough investigation of self-reference and scrupulous dwelling into formal systems, the author questions even the meaning of 'meaning', the core idea and essence of what is meaning. As Hofstadter himself wrote in the Preface to *GEB*'s Twentieth-anniversary Edition, 'GEB is a very personal attempt to say how it is that animate beings can come out of inanimate matter' (1999, p. 2).

Certain aspects of 'quantum music' were described in the article *Composing with quantum information: Aspects of quantum music in theory and practice* (2018) by Kim Helweg, in the 'Quanthoven' paper (*A Quantum Natural Language Processing Approach to Musical Intelligence*) by Eduardo Reck Miranda, Richie Yeung, Anna Pearson, Konstantinos Meichanetzidis, Bob Coecke, and in the article 'Towards the Intuitive Understanding of Quantum World: Sonification of Rabi Oscillations, Wigner functions, and Quantum Simulators' by Reiko Yamada, Eloy Piñol, Samuele Grandi, Jakub Zakrzewski, Maciej Lewenstein.

Further information about Pythagorean understanding of intervals, mathematical representation of rhythmic systems of notation and others aspects of relations between music and mathematics can be found in the works of John Fauvel, Raymond Flood and Robin Wilson *Music and Mathematics: From Pythagoras to Fractals* (2006); Gareth E. Roberts *From Music to Mathematics: Exploring the Connections* (2016); Alexander Voloshinov *Matematika i Iskusstvo* (Mathematics and Art) (2000) and others. In his book *Music and the Making of Modern Science* (2014), Peter Pesic investigated the role of music and its interaction with physics throughout the history of the development of modern science; Harvey E. White and Donald H. White mostly dedicated their book *Physics and Music: The Science of Musical Sound* (2014) to acoustical phenomena and the nature of sound.

Kinko Tsuji and Stefan C. Müller in their book *Physics and Music: Essential Connections and Illuminating Excursions* (2022) provided encyclopedic exploration of interconnections between physics and music. They start with a discussion about the physical characteristics and properties of sound, then go to tonal systems and some elementary music theory, covering the basics of notation. This is followed by explaining intervals, their roles and representations, tuning and different tuning systems across the world and throughout history, and it leads to acoustics, covering harmonics, timbre and other aspects. Each of these sections is supported by mathematical evidence describing physical phenomena. Interestingly, authors compare European systems (concepts, nature, philosophy and theory of music) with non-European systems and world music. A special case is the deep study of instruments: Tsuji and Muller not only elucidated the individual features of different instruments, but also provided rigorous classification, acoustic properties and overall succinct yet profound information about each group and single instrument in question. The explanation of the system of functioning of each instrument is exceptionally important study that can be useful for many composers. This is developed into a discussion about genres in classical music and instrumental groups. Research on acoustics (including Fourier analysis and other mathematical apparatus) penetrates every corner of this book, which is a fascinating approach that deepens the reader's understanding of how physical processes underlie acoustics and music making. Finally, the last chapter discusses aspects of hearing, sound and music perception as well as music psychology and even the sociology of music.

Andrew May in *The Science of Music: How Technology Has Shaped the Evolution of an Artform* (2023) begins with a discussion of the physics of sound, sound waves, the power of wave and decibels, acoustics, Fourier analysis, and then writes on musical parameters, algorithmic composition, music evolution, music works inspired by science and cooperation between musicians and physicists (like in case of a drummer Mickey Hart and a Nobel Prize-winning astrophysicist George Smoot). He describes the influence of the development of technologies on the world of music, from Johann Maelzel's self-contained mechanical orchestra *Panharmonicon* to *Ondes Martenot*, *Thereminvox* and contemporary OpenMusic developed by IRCAM, signal processing and even AI. Additionally, May mentions musical works written by computers (*Illiac Suite* and *Hello World*) and compares them with music written by humans (e.g. I. Xenakis *ST/4–1,080,262* for string quartet), touching on the crucial differences and providing his view and answer on why human-written music retains its validity and uniqueness. Interestingly, May states that the work of

Karlheinz Stockhausen has not only been influential among classical musicians, but also had a great impact on popular music artists such as Paul McCartney and the Beatles, Frank Zappa, Miles Davis, Pink Floyd, David Bowie and many many others. In fact, Paul McCartney studied the works of Stockhausen, Varèse and Berio, so many of the effects popularised by the Beatles were originally inventions of Stockhausen (pp. 102–103).

Among other notable literature in a given direction we can mention important works by Stephon Alexander *The Jazz of Physics: The Secret Link Between Music and the Structure of the Universe* (2016), Juan G. Roederer *The Physics and Psychophysics of Music: An Introduction* (2008), David Sulzer *Music, Math, and Mind: The Physics and Neuroscience of Music* (2021), Michael Edgeworth McIntyre *Science, Music, and Mathematics: The Deepest Connections* (2021), Godfried Theodore Patrick Toussaint *The Geometry of Musical Rhythm: What Makes a "Good" Rhythm Good* (2019), James S. Walker and Gary W. Don *Mathematics and Music: Composition, Perception, and Performance* (2019), David Wright *Mathematics and Music (Mathematical World)* (2009), Jack Douthett, Martha M. Hyde, Charles J. Smith *Music Theory and Mathematics: Chords, Collections, and Transformations (Eastman Studies in Music)* (2008), Myles W. Jackson *Harmonious Triads: Physicists, Musicians and Instrument Makers in Nineteenth-Century Germany* (2006), Cris Forster *Musical Mathematics: On the Art and Science of Acoustic Instruments* (2010), Maria Mannone *Mathematics, Nature, Art* (2019), *Simmetrie fra Matematica e Musica—I Quaderni di Clio* (2020) translated by Federico Favali and Maria Mannone, *Mathematical Music Theory: Algebraic, Geometric, Combinatorial, Topological and Applied Approaches to Understanding Musical Phenomena* (2018) edited by Mariana Montiel and Robert W Peck, and many other works by a large number of researchers.

Touching music's relation with mathematics and geometry, certain elements of dodecaphony have been labelled as set theory in music, while set theory also exists in mathematics. Others have already done a variety of research in this fascinating field. For example, Guerino Mazzola has developed category (topos) theory and applied it to music theory (2017, 2022), and Dmitri Tymoczko wrote on *A Geometry of Music* (2011) and *The Generalized Tonnetz* theory (2012). Italian composer and theorist Carmine Emanuele Cella wrote on logical and algebraic approaches to music as well as clustering techniques in his doctoral dissertation *On symbolic representations of music* (2011), and developed substantial works in the direction of computational musicology, musical informatics, sound analysis and the role of new technology in music, including his software *Orchidea* for computer-assisted orchestration (2022). Davorin Kempf's research on the basics of symmetry and variation provides numerous connections with the world of philosophy, mathematics, sciences, and especially quantum mechanics, as shown in his doctoral dissertation *Symmetrie und Variation als kompositorische Prinzipien. Interdisziplinäre Aspekte* (2006). The contemporary British composer Emily Howard used some elements of mathematical approach and computation in her music compositions, for example in *Four Musical Proofs and a Conjecture* for string quartet (2017), which was her collaboration with mathematician Oxford Professor Marcus du Sautoy. Some additional investigations of connections between mathematics/physics and classical music can be found in Massimo Blasone's articles *A Physicist's view on Chopin's Études* (2017) and *La Fisica e la Matematica negli Studi di Chopin* (2021). Aspects of connections between classical physics and classical music were researched in the article *Fractal patterns in music* (2023) by John McDonough and Andrzej Herczynski.

Biologist and physiologist Denis Noble compares systems biology (systems-level view of life) with music, asking 'where is the score and who was the composer?' in his book *The Music of Life: Biology beyond Genes* (2006). In his book D. Noble several times mentioned the importance of using metaphors (2006, pp. 16–17). Throughout the book metaphors and metaphoric comparisons/analogies with music are used to explain the systems biology approach.

It is also interesting to mention that already in the 1960s Umberto Eco attempted to correlate the emergence of the open form in art with certain phenomena of quantum mechanics, specifically, the idea of complementarity (1997, pp. 52–53, 1989, p. 15). Additionally, U. Eco highlighted certain facets of indeterminacy and uncertainty in art and analysed different forms of its implementation.

References

Abdyssagin, R.-B. 2013. *Mathematics and Contemporary Music*. Almaty: Kazak Universiteti.

Alexander, S. 2016. *The Jazz of Physics: The Secret Link Between Music and the Structure of the Universe*. New York: Basic Books.

Blasone, M. 2017. A Physicist's view on Chopin's Études. *The European Physical Journal Special Topics* 226: 2715–2728. Springer Nature.

Blasone, M. 2021. La Fisica e la Matematica negli Studi di Chopin. *Ithaca: Viaggio nella Scienza* XVII.

Cantor, G. 1874. Über eine Eigenschaft des Inbegriffes aller reellen algebraischen Zahlen. *Journal Für Die Reine und Angewandte Mathematik* 77: 258–262.

Cella, C.-E. 2011. *On Symbolic Representations of Music*. PhD dissertation, University of Bologna.

Cella, C.-E. 2022. Orchidea: A Comprehensive Framework for Target-based Computer-assisted Dynamic Orchestration. *Journal of New Music Research*.

Descartes, R. 1650. *Musicae Compendium*. Utrecht: Trajectum ad Rhenum.

Douthett, J., M.M. Hyde, and C.J. Smith, eds. 2008. *Music Theory and Mathematics: Chords, Collections, and Transformations (Eastman Studies in Music)*. Boydell & Brewer: University of Rochester Press.

Eco, U. 1989. *The Open Work*. Translated by A.Cancogni. Cambridge, MA: Harvard University Press.

Eco, U. 1997. *Opera Aperta*. Milan: RCS Libri S.p.A.

Euler, L. 1727. *Dissertatio physica de sono*. Basel: E. & J.R.Thurnisiorum.

Euler, L. 1739. *Tentamen novae theoriae musicae*. St. Petersburg: Typographia Academiae Scientiarum.

Fauvel, J., R. Flood, and R. Wilson. 2006. *Music and Mathematics: From Pythagoras to Fractals*. Oxford: Oxford University Press.

Forster, C. 2010. *Musical Mathematics: On the Art and Science of Acoustic Instruments*. San Francisco: Chronicle Books.

Heisenberg, W. 2000. *Physics and Philosophy*. Penguin Classics.

Helweg, K. 2018. Composing with Quantum Information: Aspects of Quantum Music in Theory and Practice. *Muzikologija*. 61–77. https://doi.org/10.2298/MUZ1824061H.

Hofstadter, D. 1999. *Gödel, Escher, Bach: an Eternal Golden Braid*. Twentieth-anniversary Edition. New York: Basic Books.

Howard, E. 2017. *Four Musical Proofs and a Conjecture* for String Quartet. Edition Peters.

Jackson, M.W. 2006. *Harmonious Triads: Physicists, Musicians and Instrument Makers in Nineteenth-Century Germany*. Cambridge, MA: MIT Press.

Kempf, D. 2006. *Symmetrie und Variation als kompositorische Prinzipien. Interdisziplinäre Aspekte* (Symmetry and Variation as Compositional Principles. Interdisciplinary Aspects). Dr. phil. inauguraldissertation. Freie Universität Berlin.

Kepler, J. 1619. *Harmonices Mundi*. Linz.

Mannone, M. 2019. *Mathematics, Nature, Art*. Palermo: Palermo University Press.

Mannone, M., F. Favali, and translators. 2020. *Simmetrie fra Matematica e Musica— I Quaderni di Clio*. Palermo: Palermo University Press.

May, A. 2023. *The Science of Music: How Technology Has Shaped the Evolution of an Artform*. London: Icon Books Ltd., Hot Science series.

Mazzola, G. 2017. *The Topos of Music*. Second Edition. Springer Nature.

Mazzola, G. 2022. *Functorial Semiotics for Creativity in Music and Mathematics*. Springer Nature.

McDonough, J., and A. Herczyński, 2023. Fractal Patterns in Music. *Chaos, Solitons and Fractals* 170. 113315. Elsevier.

McIntyre, M.E. 2021. *Science, Music, and Mathematics: The Deepest Connections*. World Scientific Publishing Company.

Mersenne, M. 1636. *Harmonie universelle, contenant la théorie et la pratique de la musique*. Paris: Chez Sébastien Cramoisy, Imprimerie Royale.

Miranda, E.R., R. Yeung, A. Pearson, K. Meichanetzidis, and B. Coecke. 2021. A Quantum Natural Language Processing Approach to Musical Intelligence. arXiv: 2111.06741.

Montiel, M., and R.W. Peck, eds. 2018. *Mathematical Music Theory: Algebraic, Geometric, Combinatorial*. Topological and Applied Approaches to Understanding Musical Phenomena: World Scientific Publishing Company.

Noble, D. 2006. *The Music of Life: Biology beyond Genes*. Oxford and New York: Oxford University Press.

Pesic, P. 2014. *Music and the Making of Modern Science*. Cambridge, MA: MIT Press.

Roberts, G.E. 2016. *From Music to Mathematics: Exploring the Connections*. Baltimore: Johns Hopkins University Press.

Roederer, J.G. 2008. *The Physics and Psychophysics of Music: An Introduction*. Springer Science & Business Media.

Stewart, I. 2013. *17 Equations That Changed the World*. London: Profile Books.

Sulzer, D. 2021. *Music, Math, and Mind: The Physics and Neuroscience of Music*. New York: Columbia University Press.

Taruskin, R. 2010. *Music in The Late Twentieth Century: The Oxford History of Western Music*. Oxford and New York: Oxford University Press.

Toussaint, G.T. 2019. *The Geometry of Musical Rhythm: What Makes a "Good" Rhythm Good?* Second Edition. Chapman & Hall/CRC Press.

Tsuji, K., and S.C. Müller. 2022. *Physics and Music: Essential Connections and Illuminating Excursions*. Springer Nature.

Tymoczko, D. 2011. *A Geometry of Music: Harmony and Counterpoint in the Extended Common Practice*. New York: Oxford University Press.

Tymoczko, D. 2012. The Generalized Tonnetz. *Journal of Music Theory* 56: 1.

Voloshinov, A. (2000) *Matematika i Iskusstvo* (Mathematics and Art). Second Edition. Moscow: Prosveshchenie.

Walker, J.S., and G.W. Don. 2019. *Mathematics and Music: Composition, Perception, and Performance*. Second Edition. Chapman & Hall/CRC Press.

Weinberg, S. 2016. *To Explain the World: The Discovery of Modern Science*. Penguin Books.

White, H.E., and D.H. White. 2014. *Physics and Music: The Science of Musical Sound*. Dover Books on Physics.

Wright, D. 2009. *Mathematics and Music (Mathematical World)*. American Mathematical Society.

Xenakis, I. 1953/1954. *Metastaseis^B*. Boosey & Hawkes.

Xenakis, I. 1992. *Formalized Music: Thought and Mathematics in Composition*. Revised Edition. Additional material compiled and edited by Sharon Kanach. Harmonologia Series No. 6. Stuyvesant, New York: Pendragon Press.

Yamada R., E. Piñol, S. Grandi, J. Zakrzewski, and M. Lewenstein. 2023. Towards the Intuitive Understanding of Quantum World: Sonification of Rabi Oscillations, Wigner functions, and Quantum Simulators. arXiv:2311.13313v1.

4

Werner Heisenberg's Musical Universe; Meeting with Martin Heisenberg

Werner Heisenberg—in addition to being one of the greatest scientists of all time and one of the founders of quantum mechanics—was also a brilliant pianist and music was inseparable part of his life.

I was interested to know more about music and its role in the life of Werner Heisenberg, so I sent an email to his son Martin Heisenberg (I found his email address on the website of the university where he taught). I mentioned that soon I might come to Germany, and would be honoured to meet him. I was happy to receive his reply where he wrote the possible dates for meeting and invited me to visit his house. Right after that I arranged my trip to Germany especially to meet him.

Martin Heisenberg (born on 07.08.1940) is a neurobiologist and geneticist, and is considered to be the founder of neurogenetics in Germany. He was a Professor at the Julius Maximilian University of Würzburg. His work has focused on researching the neurogenetics and brain of Drosophila (the fruit fly). Frau Apollonia, Martin's wife, is Countess (Grafin) von Eulenburg, thus she is representative of German nobility and aristocracy. She is a niece of physicist and philosopher Carl Friedrich von Weizsäcker, who was a student and closest friend of Werner Heisenberg, and Richard von Weizsäcker, who was President of Germany from 1984 to 1994.

Martin and Apollonia Heisenberg live in one of the wings of Schloss Reichenberg, a medieval castle near Würzburg, the city where Werner Heisenberg was born in 1901. Schloss Reichenberg is a beautiful castle surrounded by nature and green forests. It was built in the thirteenth century, and since

© The Author(s), under exclusive license to Springer Nature Switzerland AG 2024
R.-B. Abdyssagin, *Quantum Mechanics and Avant-Garde Music*,
https://doi.org/10.1007/978-3-031-63161-0_4

1376 has been owned by the family of Wolffskeel von Reichenberg. So nowadays this castle belongs to the same family which has been owning it for more than six centuries.

I came to Würzburg a day before our meeting, especially to see Werner Heisenberg's birth town. It is located in the north of Bavaria, the Main River flows through this city and it takes just an hour by train to reach it from Frankfurt am Main. While the city is definitely not a large one, it is affiliated with 15 Nobel Prize winners (mainly through the University of Würzburg), for example, as already mentioned, W. Heisenberg was born there.

Apart from that, Wilhelm Röntgen became the director of the Institute of Physics at Würzburg in 1888, and was elected the Rector of the University of Würzburg in 1893. And on 8th November 1895 W. Röntgen discovered X-rays (Röntgen rays) in his laboratory in the Würzburg Physical Institute. Currently one of the main streets in Würzburg is named in his honour—Röntgenring, and there is a special monument in honour of this epochal discovery as well as a large inscription on the building where this discovery was made. In 1901 W. Röntgen became the first ever person to receive the Nobel Prize in Physics.

Würzburg is a very beautiful city, with a refined architecture, exquisite Residenz (Palace), splendid Käppele (with a remarkable pipe organ), magnificent Marienberg Fortress which provides wonderful views of the city. In the end of World War II, in 1945, the whole city of Würzburg (its 90%) was completely destroyed and obliterated by the bombing of Allied forces. And after the war the city was completely restored. Therefore, almost everything we now see in Würzburg was rebuilt after the war.

On 13th May 2023 I arrived at 12:00 in front of the Reichenberg Schloss, as was agreed. I was welcomed by Martin Heisenberg and Frau Apollonia at the entrance, and then we went to the living room. I gave them my CD *Rakhat-Bi Abdyssagin plays Chopin and Liszt* (published by Kreuzberg records, Germany, 2021). They told me that they have listened to the video of my performance of my composition *The Will to Live* for piano and symphony orchestra (2015).

From the beginning we had an enjoyable and engaging conversation about music and beyond. For example, Frau Apollonia mentioned that there are a lot of Germans from Kazakhstan who currently live in Würzburg and in the Central Germany. These Germans were born in Kazakhstan during Soviet times, and emigrated to Germany after the Fall of Berlin Wall in 1989 and after the Dissolution of the USSR in 1991. Mostly they remained in Germany, but some of them are going back to Kazakhstan nowadays. As Frau Apollonia said, these Germans from Kazakhstan are called 'Kazakhs' by locals, and they still share patriotic feelings about Kazakhstan. I was pleased that Herr Martin and Frau Apollonia are well informed about my country, for example, they knew that Astana is a very young capital built by a strong-willed decision of the First President of Kazakhstan. We also discussed classical music, Richard Wagner's works and political views, the development of music in baroque, classicism and romanticism eras, and many other topics.

After that we went to the kitchen to have a lunch. There our discussion continued. I presented some of my ideas about metaphoric correlations between quantum mechanics and avant-garde music, in particular, I started the conversation with reflections on performance, experiment, notation, mathematics, the Copenhagen Interpretation of quantum mechanics etc. Herr Martin and Frau Apollonia showed interest in my research. During the lunch I learned a lot of new information about the exceptional role of music in the life of Werner Heisenberg.

I was already aware that in his youth Werner Heisenberg had to decide between pursuing the path of concert pianist and musician or the path of physics and scientific research, and that for the Nobel Prize money Heisenberg bought Blüthner Grand Piano because he loved its mellow sound, as was written in Barbara Blum's essay *Heisenberg and Music*. But what I was totally surprised at is that in Heisenberg family everyone was attached to music. Werner himself was a skilled pianist, and his wife Elisabeth sang in a beautiful soprano voice. Actually, they met in a concert of classical music on 28.01.1937 where Werner was playing piano part in Ludwig van Beethoven's Piano Trio Op.1 №2. So again it was the music that played crucial role and united Werner Heisenberg and Elisabeth Schumacher. As Herr Martin informed me, all of Werner and Elisabeth's 7 children started with singing and playing blockflöte, and then chose other instruments. Here is a list of all 7 children of Werner and Elisabeth Heisenberg and the instruments they played:

Anna-Maria Hirsch-Heisenberg (born 1938)—Piano
Wolfgang Heisenberg (1938–1994)—Violin
Jochen Heisenberg (born 1939)—Cello
Martin Heisenberg (born 1940)—Flute
Barbara Blum (born 1942)—Violin
Christine Mann (born 1944)—Viola
Verena Richert (born 1950)—Piano

According to Herr Martin, for W. Heisenberg's children there was a rule that 'you may not be a scientist, but you must be a musician'! As Herr Martin said: 'In our family it was not a question whether to learn and play music or not. It was obligatory to play music. Everyone from childhood had to learn to play music instrument, and the only question was which instrument to play. Heisenberg's children could decide whether to become a scientist or not, but playing music was a rule'.

Then, during the lunch, we had a discussion about contemporary music in Europe and in Germany in particular. I was pleasantly surprised to know that Herr Martin and Frau Apollonia are well informed about what is happening in the world of contemporary, avant-garde and academic music. For example, Frau Apollonia said that 'Karlheinz Stockhausen is already a classic. And Pierre Boulez was already an established composer in the 1950s'.

We exchanged opinions on the works of Helmut Lachenmann, Wolfgang Rihm, and came to the topic of contemporary operas and music theatre. I have spoken a little bit about my own operas: monumental *The Path Lit by the Sun* (8 acts, 27 characters, ballet, chorus and symphony orchestra, completed in 2019) which tells the story of the dissolution of the USSR and how Kazakhstan received Independence, chamber opera *The Mysterious Lady* (completed in 2019, based on Alexander Blok's drama), and Cathedral Opera *The Bruce* (2023) for soloists, reciter, choir and pipe organ (the score is published by Verlag Neue Musik Berlin), based on John Barbour's eponymous medieval narrative poem in Early Scots language (written c. 1375). In 2024 this cathedral opera had its world premieres in landmark venues such as Glasgow Cathedral (17.02.2024), St Giles' Cathedral Edinburgh (21.02.2024), the University of St Andrews, St Salvator's Chapel (24.02.2024) and Dunfermline Abbey (03.03.2024).

For me it was enriching and very interesting to hear about scientific life in Germany during World War I and World War II. Frau Apollonia told me that the situations during WWI and WWII were absolutely different in terms of intellectual life, so these two wars had a drastic contrast between each other from an academic point of view. Despite catastrophic social and economic conditions, unthinkable inflations and unrest among the people and overall unstable situation, during WWI research and work on physics and science never stopped, and immediately after WWI (during the Weimar Republic) there was a certain kind of flourishing of intellectual life. But during WWII everything stopped, and after WWII there was almost a desolation. It took a long time for Germany to recover and revive its intellectual culture after WWII and the Nazi era. Herr Martin added that until the beginning of WWII the German language was the main academic language, so scholars and researchers from all over the world used it for the communication and dissemination of ideas and novelties. There were some shifts after WWI, but

Fig. 4.1 Werner Heisenberg in the 1970s (Martin Heisenberg's private archive)

these changes were not prominent at all. Not only did the major changes, but the overall situation in the world became different after WWII. Many bright minds emigrated to the USA, so after WWII the English language started to dominate scientific communications.

After the lunch we went back to the living room. I opened my laptop and demonstrated some parts of my research on quantum mechanics and avant-garde music. In particular, I displayed presence of metaphoric analogies of Werner Heisenberg's uncertainty principle in Karlheinz Stockhausen's *Klavierstück XI* and John Cage's *Solo for Sliding Trombone*. Additionally, we discussed musical metaphors of quantum entanglement and some other quantum phenomena, and also touched ideas expressed in my sinfonietta *Ombre del Vuoto* (Shadows of the Void) for 11 performers (2019) and Symphony №2 *Chaos and Order* for full orchestra + 'audio piano' (2023). Especially we had long conversations with Frau Apollonia on that matter.

I saw Bechstein Grand Piano (it was said that this grand piano is more than 100 years old and was inherited by Frau Apollonia from her family) in the living room, and, after being granted permission, I played some of F. Chopin's works, including *Fantasie Impromptu*, selected Waltzes and Nocturnes, and excerpts from J.S. Bach's Partita №6. Herr Martin and Frau Apollonia seemed to be very touched by music, and Frau Apollonia cried sometimes during my performance. They are indeed extremely musical family with great musical taste.

After that they showed me some photos from their family archive. It was very interesting to see private photos of Werner Heisenberg. These images show that he was a shining, bright and very lively man even in his old age (Figs. 4.1 and 4.2).

Fig. 4.2 Werner Heisenberg in the 1970s (Martin Heisenberg's private archive)

It was very common for Werner Heisenberg to perform music with his children in the house, and sometimes they organised house concerts with friends. Herr Martin showed me the 'Guestbook', a small notebook where Heisenberg, his family members, friends and guests wrote down the programme of the house concerts they performed. As Herr Martin told me, there have been several such notebooks, and the one they hold dates programmes of all performances from the end of the 1940s until the death of Werner Heisenberg in 1976.

This Guestbook is very detailed, with meticulously and thoroughly written information about every house concert in great detail. Usually it lists the name of the author of the performed work, the title of the composition and the names of the performers. What is particularly fascinating is that Werner Heisenberg and his family performed concerts almost every week for almost thirty years from the end of the 1940s until 1976, as displayed in this Guestbook. The music they performed comes from a wide range of styles and epochs, including the works of J.S. Bach, G.F. Händel, W.A. Mozart, L. van Beethoven, J. Brahms and many others.

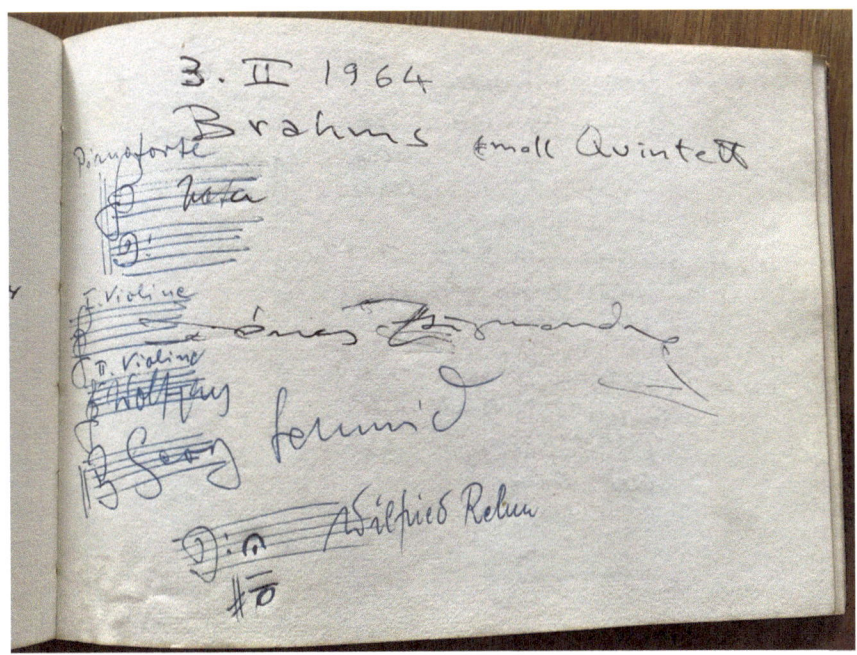

Fig. 4.3 Page from Werner Heisenberg's Guestbook with programmes of house-concerts that Heisenberg performed with his family and friends (Martin Heisenberg's private archive)

For example, a photo from the Guestbook shows that on 3rd February 1964 Heisenberg's family and his friends performed Quintett by J. Brahms (Fig. 4.3).

On 2nd and 3rd July 1966 there were house-concerts with W.A. Mozart's Piano Concerto No. 20 in D minor, K. 466 with Werner Heisenberg playing the piano solo part and Sinfonia Concertante for Violin, Viola and Orchestra in E♭ major, K. 364 (320d) with Wolfgang Heisenberg playing the violin solo and Peter Kunze performing on solo viola. Frido Mann conducted the Freundskreis-Orchester (30 musicians) with Denis Zsigmondy as Konzertmeister (Fig. 4.4).

Here are some other photos from that Guestbook. There was a concert with Mozart's music on 15th August 1966 (Fig. 4.6). All the piano parts were played by Werner Heisenberg himself (written as 'Vater' in the programme). On 31st December 1967 there was a concert of Bach's works, and Martin Heisenberg played the flute part (Fig. 4.5). Some other concerts included also Barbara (violin), Wolfgang (viola), Jochen (cello) and others. Werner Heisenberg himself always performed piano part, or the transcription of orchestral accompaniment on piano. His life looked like he was a professional pianist.

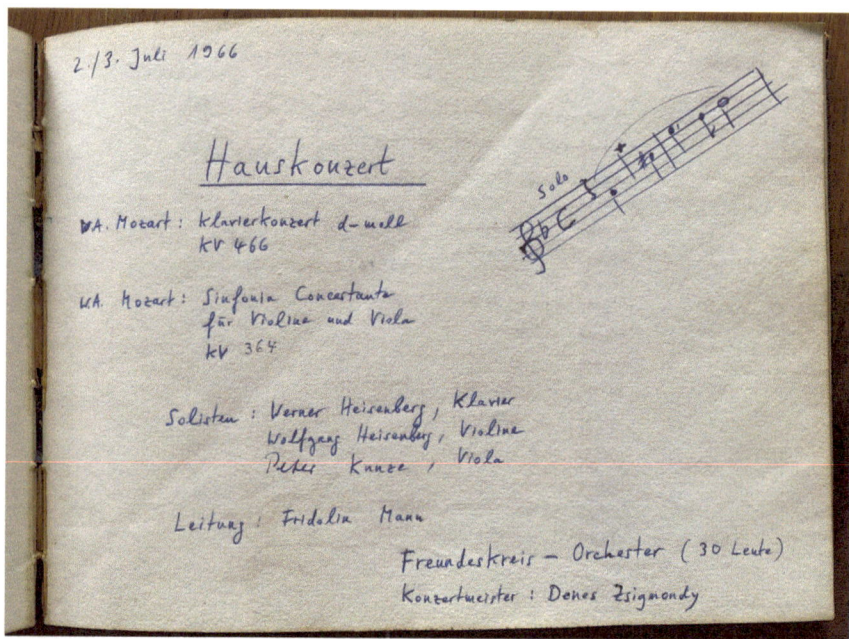

Fig. 4.4 Page from Werner Heisenberg's Guestbook with programmes of house-concerts that Heisenberg performed with his family and friends (Martin Heisenberg's private archive)

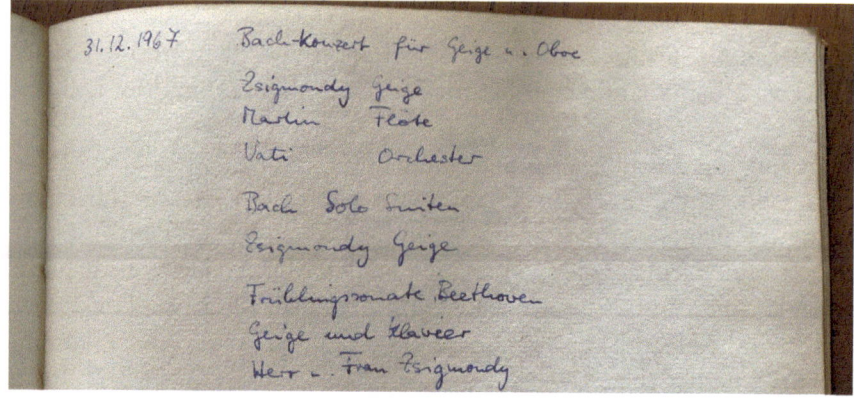

Fig. 4.5 Page from Werner Heisenberg's Guestbook with programmes of house-concerts that Heisenberg performed with his family and friends (Martin Heisenberg's private archive)

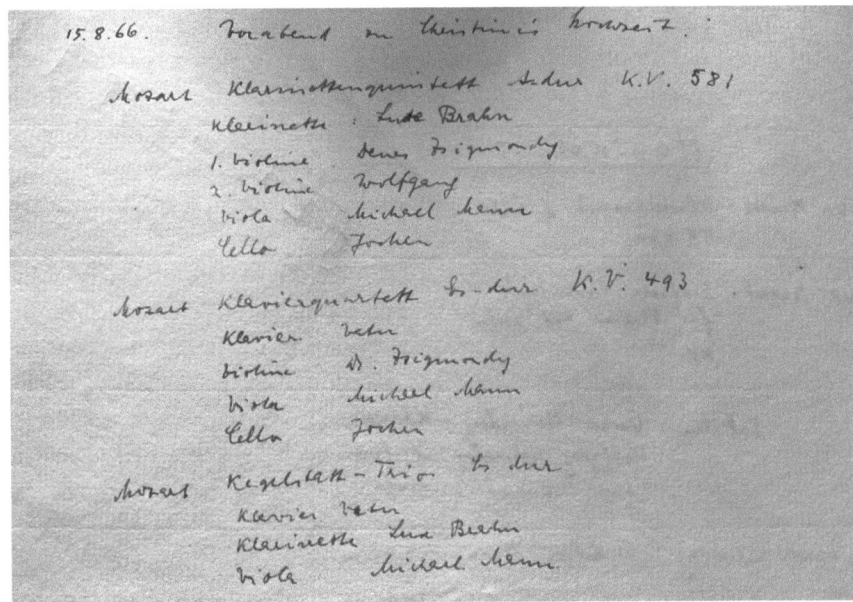

Fig. 4.6 Page from Werner Heisenberg's Guestbook with programmes of house-concerts that Heisenberg performed with his family and friends (Martin Heisenberg's private archive)

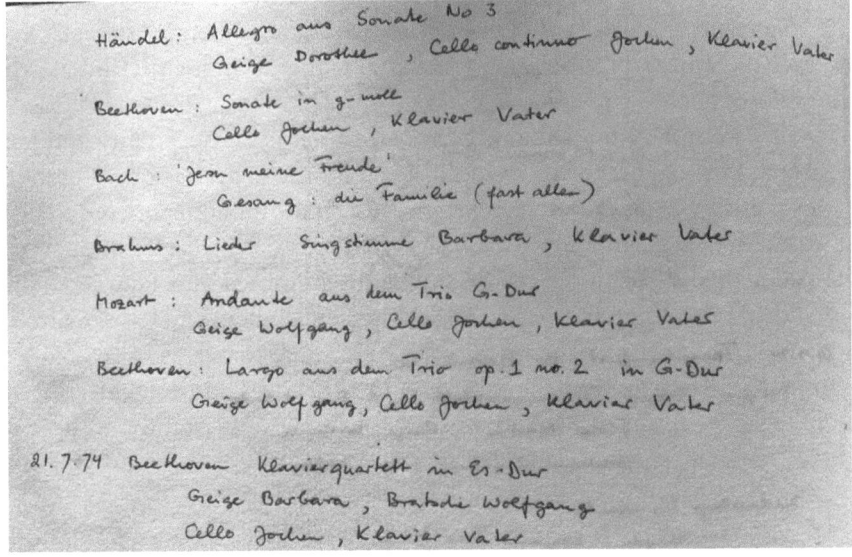

Fig. 4.7 Page from Werner Heisenberg's Guestbook with programmes of house-concerts that Heisenberg performed with his family and friends (Martin Heisenberg's private archive)

After reviewing documents and photos from the private archive of Herr Martin and Frau Apollonia, I asked Herr Martin to sign his father's book *Physics and Philosophy*—one of my desk books which I thoroughly read with great interest. I said: 'If you father were alive now, I would ask him for autograph. But unfortunately, Werner Heisenberg is not alive. So I ask you to sign for me his book'. And I was happy that Herr Martin signed for me this precious book.

Another very interesting story was told by Herr Martin about how Carl Friedrich von Weizsäcker met Werner Heisenberg and decided to become a physicist. In the 1920s Weizsäcker family resided in Copenhagen, Denmark, because the head of the family worked as a diplomat in the German Embassy. And they (including Carl Friedrich) went to one event where Werner Heisenberg gave a talk and performed a concert. There young Carl Friedrich approached Werner Heisenberg (whose scientific works he knew) and said that he wanted to become a philosopher.

To this Werner Heisenberg responded that one cannot immerse oneself in philosophy right away, first, it is necessary to study the most fundamental science—physics—in order to have 'a solid ground under the feet', and only after learning physics can one go to philosophy and other disciplines. This meeting was decisively important and fateful for young Carl Friedrich. He went to study physics under the supervision of Werner Heisenberg at Leipzig University and became a lifelong friend with Heisenberg. After all, Carl Friedrich became a prominent physicists and highly influential intellectual and philosopher.

Finally, around 18:00, I thanked hosts for their great hospitality and kind attention to me, and asked them to call a taxi for me to get back to Würzburg. At this point I was absolutely shocked because Frau Apollonia offered to drive me to the hotel where I stopped. I protested and declined this offer several times, but they insisted. So I was immensely honoured and Frau Apollonia herself drove the car all the way from Schloss Reichenberg and left me directly in front of the hotel. I again expressed my gratitude to her.

Indeed, this was one of the most memorable and significant meetings in my life (Figs. 4.8 and 4.9). And I began to closely understand and perceive Werner Heisenberg's musical universe.

Fig. 4.9 Rakhat-Bi Abdyssagin with Martin and Apollonia Heisenberg (Photo: Rakhat-Bi Abdyssagin's private archive)

Fig. 4.8 Martin Heisenberg and Rakhat-Bi Abdyssagin (Photo: Rakhat-Bi Abdyssagin's private archive)

5

Werner Heisenberg's Musical Universe; Meeting with Christine Mann

Sometime after my meeting with Martin Heisenberg and Frau Apollonia, I wrote an email to Christine Mann, a daughter of Werner Heisenberg. Her contact details were provided to me during my visit to Schloss Reichenberg. We exchanged several emails and agreed to meet in Heisenberg's family house in Urfeld am Walchensee, Oberbayern.

Christine Mann is a writer, psychologist, pedagogue and theologian. Her husband is Frido Mann, a grandson of famous writer Thomas Mann, the author of novels *Buddenbrooks* and *Doctor Faustus* in particular, and the winner of the 1929 Nobel Prize in Literature.

In 1939 Werner Heisenberg bought a house in Urfeld am Walchensee from the family of painter Lovis Corinth. The story behind this house and why Heisenbergs chose this place is revealed in *My dear Li*, a published correspondence (1937–1946) between Werner Heisenberg and his wife Elisabeth Schumacher.

Urfeld am Walchensee and the whole region including other towns such as Wallgau, Krün and surrounding areas are truly like paradise, among the most beautiful places in the world. However, it is quite complicated to reach them. But once you get there, fantastic nature embraces you so you get immersed in the wonders of the environment. Apart from being an ideal place for hiking, riding the bike and other healthy sports, it also inspires and fosters creative energy.

As agreed, I came (riding bike all the way from Wallgau) to Urfeld am Walchensee at 16:00 and found the house. Being greeted by Christine Mann and her friends, I thanked her for such a warm welcome and then we proceeded to the house (Fig. 5.1). Christine Mann looked very well, was

R.-B. Abdyssagin, *Quantum Mechanics and Avant-Garde Music*, https://doi.org/10.1007/978-3-031-63161-0_5

Fig. 5.1 Meeting with Christine Mann (Photo: Rakhat-Bi Abdyssagin's private archive)

friendly and made a good impression on me. She invited me to the table and we discussed many topics over a cup of tea. I explained the details and facets of my research dedicated to metaphoric correlations between quantum mechanics and avant-garde music, with a specific emphasis on figurative analogies of W. Heisenberg's uncertainty principle in selected works of K. Stockhausen and J. Cage. Christine Mann listened with great interest and expressed appreciation of my approach and highly evaluated the research concept.

She told me a lot about her father's relations with music. In particular, from this conversation I learned that W. Heisenberg mostly preferred composers such as J.S. Bach, G.F. Händel, F.J. Haydn, W.A. Mozart, L. van Beethoven, F.P. Schubert, R. Schumann, J. Brahms and some others. However, he did not like at all the music of G. Mahler and rejected the works of R. Wagner. He also had a cold attitude towards the symphonic music of A. Bruckner because he considered Bruckner's symphonies to be too long and pompous.

I noted that W. Heisenberg valued absolute order and harmony, which he found in the music of J.S. Bach, W.A. Mozart and L. van Beethoven. He had an acute musical taste and his mathematically and physically structured brain immediately grasped the architectonics and essence of music. For example,

Fig. 5.2 Werner Heisenberg's collection of classical music recordings (Photo: Rakhat-Bi Abdyssagin's private archive)

W. Heisenberg felt that the established tonal system began to collapse in the late works of C. Debussy and M. Ravel.

This discussion continued and Christine Mann shared a great deal of details about her father's musical philosophy and artistic preferences. Then we proceeded inside the house where she demonstrated me the collection of vinyl recordings that her father loved to listen to—this collection is full of classical music (Fig. 5.2). Another revelation displaying the evidence that classical music played exceptional role in life of W. Heisenberg and his family.

An upright piano stood in the corner of the room—a piano that Werner Heisenberg used to play! At this point Christine Mann asked me to play it. Of course I could not resist the honour of playing the piano of Werner Heisenberg himself! So I performed an array of works by F. Chopin (some waltzes, nocturnes, preludes and Fantaisie-Impromptu in C♯ minor, Op. posth. 66, WN 46), F. Schubert, J.S. Bach (excerpts from Partita No. 6 in E minor, BWV 830), L. van Beethoven (*Für Elise* Bagatelle No. 25 in A minor, WoO 59, Bia 515) and others, and it became a mini-concert (Fig. 5.3).

When I started to play W.A. Mozart's Piano Sonata No. 16 in C major, K. 545, Christine Mann suddenly stood up and went outside the house to

Fig. 5.3 Rakhat-Bi Abdyssagin plays Werner Heisenberg's piano in a presence of Christine Mann (Photo: Rakhat-Bi Abdyssagin's private archive)

the terrace. When I finished playing, she returned to the room in tears and said that this performance has especially pleased and touched her, because her father (W. Heisenberg) usually played this Mozart's sonata on this piano and she as a child listened to it in the terrace, and my performance reminded her of this episode.

It was not by chance that I chose this composition of Mozart; I was aware from the published letters (*My dear Li*) that Werner Heisenberg loved to play this sonata.

After this improvised concert she showed me the music shelf of her father with scores of various classical composers (Figs. 5.4 and 5.5). These scores were acquired at almost the same time, but had varying degrees of 'being

Fig. 5.4 Classical music scores used by Werner Heisenberg (Photo: Rakhat-Bi Abdyssagin's private archive)

used', so we could easily conclude that, for example, the music of J.S. Bach, W.A. Mozart, L. van Beethoven, F.P. Schubert and R. Schumann was very often played by Werner Heisenberg, while the music of F. Chopin was rarely performed (judging from 'almost untouched' condition of his scores).

Finally, I asked her about her father's performance with the Radio Symphony Orchestra of Bavaria, and she promised me to send more information and the title of the specific piano concert by W.A. Mozart that was performed on that concert. She told that after rehearsals with professional orchestra of such a level, W. Heisenberg said that 'he had not realised before, how precise the tempi must be and how much coordinated should the hands be'.

I thanked her for her cordial hospitality and asked her to send some photos of her father on piano (Figs. 5.6 and 5.7) when she returns to her home in Göttingen.

Fig. 5.5 Classical music scores used by Werner Heisenberg (Photo: Rakhat-Bi Abdyssagin's private archive)

Fig. 5.6 Werner Heisenberg plays piano with a string trio (Heisenberg family's private archive)

Fig. 5.7 Werner Heisenberg accompanies his daughter Barbara Blum's singing (Heisenberg family's private archive)

6

Great Scientists—Gifted Musicians

The legendary physicist Albert Einstein (1879–1955), who was also a violinist himself, once said: 'Mozart's music is so pure and beautiful that I see it as a reflection of the inner beauty of the universe'.

Max Planck (1858–1947)—the physicist who discovered the 'energy quanta'—was also a gifted musician with skills of playing piano, organ and cello, as well as composing. As P. Pesic wrote, Planck had an 'acute sense of pitch' and was 'a pianist of considerable skill. As a student, he had composed songs and even an entire operetta that was performed in the musical evenings that were fixtures of professorial life in those days; he conducted choruses and orchestras, played the organ at church services, and studied harmony and counterpoint [...] After Planck returned to Berlin in 1889 as a professor, his own home music-making included Joseph Joachim and Albert Einstein. Every other week he conducted an informal chorus that included his children, neighbors, and friends' (2014, pp. 255–269).

In the 1920s Max Planck and Albert Einstein (who at that time was already in Berlin) played music together. As Prof. Konrad Kleinknecht (a physicist and co-founder of the Heisenberg Society in Munich) wrote: 'Planck, a nearly professional pianist, hosted regular chamber music sessions—sometimes with virtuosos like the violinist Joseph Joachim and sometimes with amateurs like Einstein as violinist and Planck's son Erwin as cellist—for piano trios. When famous soloists came to Berlin on their concert tours, they might wish to meet the famous Einstein. Sometimes, Einstein used the occasion to invite the musical celebrities to play chamber music with him. One of these celebrities was the Russian cellist Gregor Piatigorsky, who became principal cellist of the Berlin Philharmonic Orchestra from 1924 to 1929 under Wilhelm

R.-B. Abdyssagin, *Quantum Mechanics and Avant-Garde Music*, https://doi.org/10.1007/978-3-031-63161-0_6

Furtwängler. With Piatigorsky and a pianist, Einstein played a piano trio. On one occasion, when the piece was over, Einstein asked Piatigorsky how his playing had been. Piatigorsky replied, "Relatively good!".' (2019, p. 155).

Max Born (1882–1970)—a physicist and mathematician who influenced the development of quantum mechanics—played piano (sometimes accompanying A. Einstein who played violin). Max Born and Werner Heisenberg played together piano concertos of Wolfgang Amadeus Mozart and Ludwig van Beethoven on two pianos in Born's house in Göttingen (Blum 2020).

Werner Heisenberg (1901–1976)—a physicist known for the uncertainty principle—was a distinguished pianist with a keen interest in music. In addition to the information in Chapters 4 and 5, the following facts are presented. 'To Heisenberg music was a basis for living his life: it gave purpose and was indispensable' (Blum 2002). Throughout his career in physics, his love for music only increased and remained one of the strongest passions. He started to learn piano playing at an early age and 'music quickly became a central element in Heisenberg's life' (Carson 2014, p. 19).

What is particularly fascinating is that for the Nobel Prize money Heisenberg bought *Blüthner* Grand Piano because he loved its mellow sound. And I am exceptionally happy and deeply honoured because I had the privilege of playing exactly that Grand Piano of Heisenberg! On 4th December 2023 during my journey to Switzerland (near Geneva and CERN) I visited Barbara Blum, a daughter of Werner Heisenberg. In the house of Barbara Blum I had an honour of playing the personal *Blüthner* Grand Piano of Werner Heisenberg—this is exactly the same Grand Piano that Heisenberg bought right after receiving the Nobel Prize in Physics in 1933 for the money of this prize (Figs. 6.1 and 6.2). Although this Grand Piano is more than 90 years old, it is in perfect condition. And it was fully tuned in the morning especially for my visit. The honour of playing Werner Heisenberg's 'Nobel' Grand Piano is one of the most symbolic and significant events in my life.

One of Werner Heisenberg's ancestors, August Zeising, was a recognised violinist and a student of Louis Spohr; therefore, Heisenberg's family had a long musical tradition. Werner Heisenberg himself started to learn playing piano at the age of five. In 1932 he studied counterpoint and fugue, and even tried to compose a fugue himself (Kleinknecht 2019, p. 155).

As his daughter Christine Mann said, for his 60th birthday in 1961 Werner Heisenberg performed Wolfgang Amadeus Mozart's *Piano Concerto No. 23* A major KV 488 with the Bayerischer Rundfunk. More about his relations with music are revealed in Barbara Blum's essay *Heisenberg and Music* (2002) and Konrad Kleinknecht's book *Einstein and Heisenberg* (2019).

Fig. 6.1 Rakhat-Bi Abdyssagin plays *Blüthner* Grand Piano which Werner Heisenberg bought for the Nobel Prize in 1930s. Photo made on 4th December 2023 (Photo: Rakhat-Bi Abdyssagin's private archive)

Fig. 6.2 Rakhat-Bi Abdyssagin and Barbara Blum. 4th December 2023 (Photo: Rakhat-Bi Abdyssagin's private archive)

Considering the relations between music and mathematics, W. Heisenberg himself wrote: 'Even if one carefully analyzes other representations of reality as well, such as music or the creative arts, which are far removed from the natural sciences, they will reveal internal orders that are very closely related to mathematical laws. Those orders can be as clearly discerned as in a Bach fugue, for example, or in a symmetrical ribbon ornament; they might be noticed initially through a unique balanced quality, through the immediately evident beauty of a melodic line such as that of the famous sub-theme in the first movement of Beethoven's D major violin concerto. Closer examination always shows simple mathematical symmetries comparable to those mathematics deals with in group theory (2019, p. 30).

Further evidence of Werner Heisenberg's profound affection for music is displayed in his phrase: 'One really can't live without music. But when one listens to music, one sometimes hits on the absurd idea that life has a meaning' (Kleinknecht 2019, p. 156).

As another contemplation of mathematical structures in music, in a letter (from 6th October 1940) to his wife Elisabeth, Werner Heisenberg wrote:

My dear Li!

It is almost midnight, and I am conducting my life very unreasonably—so it is time that you soon should come here again. I have studied the E-flat-major concerto for almost two hours, and within me I am quite caught up and excited by the glorious music even now. Overnight, I have mentally constructed a Corona talk about this concerto—I was not sleeping well—with an exact analysis of the architecture, the mathematics in the different motifs, a demonstration on the piano of different themes with their instrumentation and, in the end, the entire concerto ought to be performed. The work is so incredibly complete that one could not do justice to the sheer number of symmetrical interrelationships in the architecture. But unfortunately, I do not have such a Corona talk in my immediate future. (Werner and Elisabeth Heisenberg 2016, p. 134)

According to W. Heisenberg's daughters Christine Mann and Anna-Maria Hirsch-Heisenberg, 'E-flat-major concerto' (mentioned in the letter) is Piano Concerto No. 5 in E-flat major Op. 73 *Emperor Concerto* by Ludwig van Beethoven. Thus, this is the piano concerto in which Werner Heisenberg 'discovered all the mathematical construction' (Christine Mann's email to Rakhat-Bi Abdyssagin, 10 August 2023).

A 'Corona talk' presumably refers to the circle of Leipzig professors called 'Coronella' in which W. Heisenberg took part. Helmut Rechenberg provided more details of this 'Coronella' group in July 1988 in his introduction to

W. Heisenberg's *Ordnung der Wirklichkeit*: 'A small circle of professors at the University of Leipzig that Heisenberg belonged to allowed relaxed and informative conversations on matters that went beyond the narrower area of their special disciplines. The circle was called 'Coronella'; in addition to Heisenberg there was the art historian Theodor Hetzer, the Nordic linguist Konstantin Reichhardt, the historians Helmut Berve and Hermann Heimpel, the archeologist Bernhard Schweitzer and the classical philologists Friedrich Klingner and Wolfgang Schadewaldt. They met regularly in private sessions where they presented their lectures, followed by informal discussion. The lectures and addresses they gave in public either corresponded to those in their private circle or had had their 'dress rehearsal' there. Later Heisenberg maintained his relationship with the members of 'Coronella' over many years' (2019, p. 9).

Victor Frederick Weisskopf (1908–2002)—was a distinguished theoretical physicist who served from 1961 to 1966 as the director-general of CERN, and liked to say that 'his favorite occupations were Mozart and quantum mechanics' (Gottfried and Jackson 2003).

Alfred Brian Pippard (1920–2008)—was a British physicist based in Cambridge and Cavendish Professor of Physics from 1971 until 1982. He was also a 'very fine pianist' (according to Andrew Briggs), and even 'considered for a time a career in music: his piano-playing was of concert standard' (Waldram 2008). It is known that 'music played a central role in Brian's life. He began learning the piano at the age of five years but it was only at Clifton Preparatory School at the age of ten years that his imagination was sparked by the great classical composers. [...] his interest in music was really serious. He made early attempts at composition, but later destroyed most of his efforts' (Longair and Waldram 2009, p. 204).

Brian David Josephson (born 04.01.1940)—a theoretical physicist who won the Nobel Prize in Physics (1973) at the age of 33 for the pioneering research on superconductivity and quantum tunnelling, for the 'Josephson effect' (which he discovered at 22 years old while working on his PhD at the University of Cambridge). The 'Josephson effect' has enormous fields of practical application, with one of the remarkable cases being the MRI (magnetic resonance imaging). Prof. Josephson is also known for his research and belief in paranormal activities and extrasensory perception. In addition, Prof. Josephson has a keen interest in music. He wrote a musical composition *Sweet and sour harmony* and penned articles (co-authored with Tethys L. Carpenter) *Musical minds* (1991), *Music and Mind—a Theory of Aesthetic Dynamics: On Self-Organization* (1994) and *What can music tell us about the nature of the mind? A Platonic model* (1996).

On 25th April 2023 I was honoured to meet for the first time with great physicist Brian David Josephson in Trinity College, University of Cambridge. I wanted to meet Prof. Josephson because I found very interesting the ideas expressed in his article (co-authored with Tethys Carpenter) *What can music tell us about the nature of the mind? A Platonic model*, the copy of which he kindly signed for me. I was surprised and at the same time pleased to know that Prof. Josephson listened to my works before meeting and he liked my symphonic poem *Qubylys* and *The Will to Live* for piano and orchestra.

To see, to visit Trinity College Cambridge itself is a great luck, and to have a meeting, a conversation with a Nobel Prize winner is a golden ticket!

I was even more surprised that the professor himself personally met me at the entrance to Trinity College and then accompanied me all the time. Trinity College was founded in 1546 by King Henry VIII Tudor and remains one of the most famous colleges of the University of Cambridge. It is possible to speak endlessly about it, but I would like to share what I was especially impressed by. Trinity College fascinates right after passing the Porter's Lodge, which opens wonderful and miraculous view to the Great Court.

Prof. Josephson invited me to Fellows' lunch at high table (special table for Fellows/Professors and their guests), and I was thrilled right after entering— by seeing the portraits of Isaac Newton, James Clerk Maxwell, Joseph John Thomson, Ernest Rutherford, Lord Rayleigh and many others who dined at this place!

In general, 121 Nobel Prize winners are affiliated with the University of Cambridge, 34 of whom studied or worked at Trinity College! Apart from already mentioned legendary names, Nobel Laureates affiliated with the college include Niels Bohr, William and Lawrence Bragg, Charles Glover Barkla, Andrew Huxley, Pyotr Kapitsa, Didier Queloz and, of course, Brian Josephson! Famous alumni of Trinity College include King Charles III, Lord Byron, Vladimir Nabokov, Ludwig Wittgenstein, Bertrand Russell, G.E. Moore and many others.

Speaking of the importance of Trinity College, its Master (head/leader of the college) is appointed by the Crown (candidate is proposed by Fellows). Current Master is Dame Sally Davies, and former Master was Gregory Winter, molecular biologist who received Nobel Prize in Chemistry!

During lunch with Prof. Josephson, he revealed that he can play four music instruments. We also discussed his music composition *Sweet and sour harmony* for voices and violins.

After the lunch, we continued our discussion in the Senior Parlour— a specially designated place for fellows/professors of the college. We had a conversation about my project on metaphoric correlations between quantum physics and avant-garde music, and it was exceptionally enriching to hear comments from Prof. Josephson. In particular, we discussed Pauli exclusion principle, Heisenberg's uncertainty principle, observer effect, proof in mathematics and in philosophy, ideas of David Bohm and some aspects of my sinfonietta *Ombre del Vuoto* (Shadows of the Void, 2019). In general, it was a substantive, very useful conversation for me.

After the discussion, he showed me inner court of the college and its magnificent chapel. When we entered the chapel, we saw the statues of Isaac Newton and other great men, and also the pipe organ. Passing under the organ, I said that Prof. Josephson's music can sound well if performed on this organ with chapel's choir. Professor said that this is an interesting idea, and wrote this idea down in his notebook.

Then we visited the park of the college—an impressive Fellows' Garden, a place almost from a fairy tale, which includes such historic areas as Newton's lawn and Newton's apple tree.

After Fellows' Garden, Prof. Josephson invited me to see the library of Trinity College—special sacred place, a home for a number of historic documents, valuable intellectual treasures of whole humanity. The College library has two sections—one for studying (Student Library), and another (Wren Library) containing ancient manuscripts and original works of great geniuses. Prof. Josephson used his Cambridge ID card to open the door, and only because of this I was able to enter this marvelous place. It was unimaginable experience to see with my eyes all these ancient books and handwritings of such significant figures as Ludwig Wittgenstein (his philosophical notebooks written in German) and others.

The peak of this journey to the library was that I personally saw the original documents of Isaac Newton himself! For example, through the protective glass I contemplated the first printed copy of Newton's *Philosophiæ Naturalis Principia Mathematica* (one of the most important and influential books in the history of civilisation) with handwritten comments by Newton (he was making corrections for the second edition), Newton's drawing instruments, his original prism, walking stick, undergraduate notebook and a lock of his hair (Fig. 6.3).

Fig. 6.3 With Brian Josephson during our first meeting on 25th April 2023. Wren Library, Trinity College, University of Cambridge (Photo: Rakhat-Bi Abdyssagin's private archive)

It is impossible to express in mere words my excitement of being able to witness these artefacts that shaped human civilisation!

I expressed my deep gratitude to Prof. Josephson for this wonderful and unforgettable meeting for me. Each new meeting with Prof. Josephson is an enriching and unique experience for me (Fig. 6.4).

Konrad Kleinknecht (born 23.04.1940)—is an experimental physicist who worked at CERN and at universities in Germany. He is the founder of the *Heisenberg-Gesellschaft* (Heisenberg Society) in Munich and was its first Chairman. He is also a violinist and even played together with Werner Heisenberg (piano).

Michael Berry (born 14.03.1941)—is a physicist, Melville Wills Professor of Physics (Emeritus) at the University of Bristol. He is known for the 'Berry phase' and the 'Berry connection and curvature'. In addition to being a distinguished physicist, he loves jazz and as a teenager bought a saxophone but gave it up after hearing Charlie Parker, before he earned later from physics that success requires persistence.

Ian Stewart (born 24.09.1945)—is a British mathematician and writer of popular science and science-fiction, Emeritus Professor of Mathematics at the University of Warwick. He is an author of more than 100 books, many of which are bestsellers about mathematics and popular science. According to

Fig. 6.4 With Brian Josephson at Candlemas Feast, Trinity College Cambridge on 2nd February 2024, my 25th birthday (Photo: Rakhat-Bi Abdyssagin's private archive)

him, approximately 2–3 million copies of his books were sold worldwide in many languages. Interestingly, as I have established, Prof. Stewart is a 'scientific' descendant of Isaac Newton by his 'mathematical' lineage (from supervisor to student): Isaac Newton → Roger Cotes → Robert Smith → Walter Taylor → Stephen Whisson → Thomas Postlethwaite → Thomas Jones → Adam Sedgwick → William Hopkins → Francis Galton → Karl Pearson → Philip Hall → Brian Hartley → Ian Stewart. Prof. Stewart plays guitar and was lead guitarist in the group called 'The Shades of Night' as an undergraduate in Cambridge. He currently owns 3 guitars (1 acoustic, 1 electric, and 1 semi-acoustic).

Tolegen Kozhamkulov (born 29.04.1946)—is a theoretical physicist, President of Kazakh Physical Society, an organiser and the first Director of the Research Institute of Experimental and Theoretical Physics in Kazakhstan, a leading specialist in quantum field theory, the theory of relativity and gravity, physics of the atomic nucleus and elementary particles, a laureate of Kazakhstan's Al-Farabi State Prize in Science and Technology (2015). He conducted research at the Ioffe Institute and at the Landau Institute for Theoretical Physics. He developed a scientific direction: stochastic quantization of gauge fields on a lattice (Kozhamkulov 1993), created a theory of perturbation of a new type on the background of a stochastic field (Migdal and Kozhamkulov 1984), and 'for the first time proved the presence of the phenomenon of confinement in lattice quantum chromodynamics (QCD)' (National Academy of Sciences of Kazakhstan 2022, pp. 111–113). In particular, 'The covariant stochastic equation in curved field space is derived' (Kozhamkulov 1985). In addition, he plays dombyra (Kazakh national instrument), bayan (accordion) and piano. He is an author of music works *Elegy of Soul* for piano solo, *Stream of Thoughts* for orchestra of Kazakh national instruments, and a song *Remember your Native Land*. He is the Honorary Professor of Kazakh National Academy of Music (2006).

Martin Chalfie (born 15.01.1947)—an American biologist famous for the introducing the use of GFP, green fluorescent protein, as a biological marker for which he was awarded the 2008 Nobel Prize in Chemistry along with Osamu Shimomura and Roger Y. Tsien. He is a University Professor at Columbia University in the City of New York.

Martin's father, Eli Chalfie, was a professional musician, and Martin plays guitar himself. During our online conversation (6 November 2023) he talked about his deep connection with music and other aspects of his family life.

As Prof. Chalfie recalls: '*My father had an ability to pick up instruments and play. He was born in 1910 and the family story is that, in his 20's, his musical talents helped his family get through the Great Depression (1929–1939), and especially allowed his younger brother to go to college and law school. At the time in the United States the big entertainment was radio. Certain large radio stations employed studio musicians to play behind all their various shows. My father was a studio musician for one of these very large radio stations, WLW* in Cincinnati. *During World War II, my father served as a musician in the US Navy, performing on at various places in the US and on his ship, the USS Missouri. However, after the WWII he didn't want to do all the traveling involved in being a musician and decided to stop playing professionally*'.

Then Martin tells how he first started to learn playing guitar: '*When I was about 12, my father gave me a Gibson C1 classical guitar and learned that a very prominent classical guitar teacher, Richard Pick, lived nearby. When my father asked Pick if he would teach his son, Pick replied that he didn't teach young children. My father than asked if Pick would teach me if my father could teach me from Pick's first instruction book. He agreed and my father was my first teacher. After a year, I had gone through the book and Pick agreed to teach me. But first he asked about my age. When my father said I was 13, Pick was surprised, saying "when I said I didn't teach young children, I meant four and five year olds". Nevertheless, I'm very grateful that I had the first year of my guitar with my father'.*

Martin goes into details of the story how his guitar practice progressed: '*I studied with Pick for about four years, basically through high school. I went to college at Harvard and played for my own enjoyment, playing a lot of the pieces that I had learned from Pick and picking up some things along the way. Just after I graduated from college, I discovered that Harvard had a wonderful microfilm library of Renaissance and Baroque lute music on the second floor of Memorial Church. Although this archive room had all these microfilms, hardly anybody knew it existed. And I would go there and hand copy various pieces of lute music and then go back to my room and play them. In addition to this repository, I discovered that the Boston Museum of Fine Arts had a room with a musical instrument collection. I would go there and play their reproduction of a Spanish lute, a vihuela, made by Donald Warnock for an hour or so. I had a wonderful time'.*

Martin said that he tries to compose music, but the speed of writing music greatly varies, and sometimes it seems that he 'will write one note per week'. Concluding, Prof. Chalfie also explained some aspects of his creative approach in music: '*There's, for me at least, a real enjoyment in just the participation in writing and playing. I guess it's a little bit like watching sports or participating in sports, I'd much rather participate than watch, I'm not a person that likes to watch. If I have some time at home, I don't turn on music to listen to it. I pick up the guitar and I play something I've either played before or just this last week, I've discovered a new site that has a lot of free sheet music. And I've been enjoying playing some of the pieces. I'm always looking for something else'.*

Hugh David Politzer (born 31.08.1949)—a theoretical physicist, currently he is Richard Chace Tolman Professor of Theoretical Physics at the California Institute of Technology. In 1973, in his first published article *Reliable Perturbative Results for Strong Interactions?* he described the phenomenon of asymptotic freedom of quarks. For the discovery of asymptotic freedom in quantum chromodynamics, in 2004 he was awarded the Nobel Prize in Physics, shared with David Gross and Frank Wilczek. In addition to making important contributions to particle physics, David is a passionate musician and plays several instruments including banjo, guitar, harmonica, autoharp, recorder flute, and baroque oboe. At Caltech he teaches a seminar course to a small group of freshmen titled 'The Science of Music'. He describes how

Fig. 6.5 David Politzer with banjo (Photo from D. Politzer's private archive)

instruments work, and also concentrates on brain science and cross-cultural comparisons, all aiming to understand what music actually is (Fig. 6.5).

During our online discussion (18 November 2023), David was very kind to share the story of his relations with music and his path in science:

'I'm a person that is full of music. I sing all day, mostly to myself. Often, something connects to some song I know, and it's right there. I pick up musical instruments. I have many, as I said, not played particularly well. They're out in the house. They're out in my office. Much of the music I know is from listening to recordings or performances by professional musicians. I feel a profound debt to them for that.

It's common for little children to take their world for granted. That's how the world is, right? My parents escaped from Czechoslovakia just two months after it was invaded by Germans. They made it through connections to a boat in Poland and sailed to England, where they were during the war. They spoke two languages in the home of my father when he was growing up, and two more around the city. So he spoke four languages, like a native, except with an accent that was different from any of the individual ones. It's the city that is now the capital of Slovakia, Bratislava. Then, the Slovaks were only the third largest ethnic group. Growing up in a small town, my mother spoke Slovak. She learned English just in the few years before I was born. They came to the US after World War II. I have to credit my mother with figuring out all kinds of things about how things worked in her new life and getting over the trauma of the past. In her family there were few survivors, and the whole community was gone. Among other things, she found great opportunities in New York City for her children.

When I was very little, there was some music programme nearby. It was just rhythm and banging and a group singing a little bit. Then at elementary school I started to play a recorder flute. And I enjoyed that in a group lesson. I had a friend who started taking accordion lessons at a nearby music store, and so I wanted to do that. And this is still while in elementary school. I didn't practice much. My friends in high school all could strum a guitar. I picked that up, and I started to play the banjo. I thought it was amazing.

In seventh and eighth grade, I was in a boys choir, and I loved it. In school every lunch hour, everyone else went to a cafeteria. The choir members went to the music room and we had to have our lunch in a brown bag. We would perform in the city for free at the train station and hospitals. The only name I remember of a composer of the music that we sang is Palestrina (sixteenth century Italian composer). I sang soprano, and we did this stuff right. Then my voice changed. This was a catastrophe because I couldn't sing anymore in the choir.

In an effort to replace the singing, I bought a harmonica, and by the age of 15 played harmonica and guitar. Then I decided I'm going to build a banjo because I was still paying for the first guitar that I bought. I couldn't afford a new one. So I built it. I lived in the New York City, but it's different now. Then, there was a street with music stores. You could buy tuners, fret wire, the hooks, the hardware of a banjo etc. From furniture makers, I scavenged mahogany, rosewood, and maple. I have a decent banjo that I built. I took it to college and played it.

I went to the University of Michigan, Ann Arbor. Well, this is America. For a bachelor's degree, if you're not at an engineering school, if you're at a liberal arts and sciences school, no matter what you major in, you need a broad selection of courses. I had lots of courses in high school, which they accepted on the basis of tests as actual credit. So it's not just that I was put at a higher level, but on my transcript there were certain requirements that were already satisfied by my high school work. By the way, my high school physics class was more demanding than what we do with freshmen at Caltech.

But I also worked. I looked for a lab job right away. I don't know, I can't tell you why, but I liked making things, and I was kind of interested in physics. My older brother (and this is often the way families work) was my hero. He was studying physics, and the impression I got was that the really smart guys do physics. I'm six, seven years old, and I believe that. So when it comes to my time, those are the courses I choose in high school.

On arrival at college, I went to physics labs, just knocking on doors. I'm good with my hands. I can build and repair things. Can I get a job? I got a job right away. And I worked probably 15 h a week during school in a lab, and maybe 30 h a week in the summers in Ann Arbor. I learned more and more physics in labs by doing. So my background as a theoretical physicist is rather unusual, because working in the labs was what attracted me again and again to physics. But I also had built a banjo and learned to play the diatonic harmonica. I would go almost every week to live performances of traveling musicians at a folk venue, which really inspired the harmonica, guitar, and banjo playing. Today I play a five-string banjo, in a particular style and favorite tuning.

There were a lot of years when I didn't play much music, very immersed in physics. I guess that theoretical physics captured me. It started because as an undergrad, I was working with a professor planning an experiment. He had the idea, and he told me what to do. That was my job. And he picked me because I was working on other experiments of his. I guess he decided I was good at it. I didn't know I was good. I just did what I did. But when I asked questions, I realized he didn't understand much about the experiment we were planning. He understood the goal. But he didn't seem to understand why it was worth doing. I asked him. Instead of answering, he sent me to a theory professor in particle physics down the hall. And he said, 'ask this guy in New York; he could explain it to you'. I was not impressed. I decided I should learn more physics. Three of the four courses I took when I started graduate school at Harvard were in particle physics, taught by great theoretical physicists. And it was a very exciting time in particle physics, very unlike today (which is why I haven't done any particle physics in 25 years). It's very different now. At that time, there was a huge accumulation of experimental facts. Experimental data. Interesting, complicated. And there was a huge accumulation of theoretical ideas to address the facts that were mutually contradictory, and none of them particularly worked well. That's when I entered graduate school. And by the time I finished, and I made some contribution to it… but I don't know, I was in the right place right time and paid attention. Paying attention apparently is important.

By the time I finished graduate school, there was what we call the Standard Model. It's an awful name except it was chosen because in those intervening years there were a lot of other alternatives. Much more creative. It's very… What's the right word? Pedantic, no? Mundane. The model is as simple as can be and used the old tools and language but in a totally different way. That was the surprise. The equations, the Lagrangian or Hamiltonian, the forces, equations for the forces existed in the literature and in the particle physics literature. But how to use them and put them together was different. And that took place over a small number of years. And so that was the difference after. Anyway, there was a lot to do in theoretical physics, and that's what I did.

Some years later, I got seriously back into music. I suspect that the enthusiasm of my second son, then a toddler, had a lot to do with it.'

Frank Anthony Wilczek (born 15.05.1951)—is an American theoretical physicist, Herman Feshbach Professor of Physics at the Massachusetts Institute of Technology. He was awarded the 2004 Nobel Prize in Physics 'for the discovery of asymptotic freedom in the theory of the strong interaction', together with David Gross and H. David Politzer. Frank Wilczek loves classical music, plays piano and accordion, and also said that 'Bach is a great manipulator of patterns' (Nobel Foundation 2019). In a book *Fundamentals: Ten Keys to Reality* (2021) Wilczek even made some metaphoric correlation between music and physics: 'Harmony is a local analysis—here monitoring a moment in time, rather than a point in space—while melody is a more global analysis. Harmony is like position, while melody is like velocity' (p. 209).

David Deutsch (born 18.05.1953)—a physicist at the University of Oxford, one of the founders of quantum theory of computation and constructor theory, the author of the bestselling popular science book *The Fabric of Reality* (1997) and *The Beginning of Infinity* (2011)—plays piano and admires classical music.

As he himself recalls (at our online meeting on 6 November 2023), '*I enjoy playing the piano and I have been trying for years to learn how to play the Wolfgang Amadeus* Mozart's Rondo in A minor KV 511 (Fig. 6.6). *And there are various things that prevent me from learning it properly. One is that as soon as I'm playing it, I don't want to stop. So I can't stop to correct an error. I have to go right through to the end. And then I've forgotten all the errors and I've forgotten all the fingering that I should get right and everything. And so that's one reason. Another reason is that… I can't play it well. So I have to play it in a way which to me evokes the right sensations. Even though if a professional was playing it, I would want them to play it differently from the way I play it. This Mozart's* Rondo in A minor KV 511 *is not as difficult as Ludwig van Beethoven's sonatas, for example* Piano Sonata No. 32 in C minor, Op. 111, Piano Sonata No. 31 in A flat

Fig. 6.6 First page of the manuscript of W.A. Mozart *Rondo in A minor KV 511* (Public domain. *Source* IMSLP, Petrucci Music Library)

major, Op. 110 or Piano Sonata No. 23 in F minor, Op. 57 Appassionata. *It's less difficult than those, but there's something in it which transcends those in my view. Maybe I'm just old-fashioned. I'm from the eighteenth century, not nineteenth. Beethoven is fantastic. It's wonderful. But he achieves that by exaggerating things, by making things loud, by making things fast and so on. Whereas Mozart achieves the same level of sublime music by just using the ordinary alphabet of music as it existed in his century*'.

Then David goes into the story how his introduction to the world of music began: '*I started when I was a child. The reason I started playing the piano was that my primary school was a really bad and boring, tedious place. So I signed up to music classes so that from time to time I would be randomly taken out of the lesson and to do something less boring. When I left that school and went to my next school, I was already hooked enough to want to continue lessons. So I asked my parents to find me a music teacher, but that music teacher was terrible. For a start, she only wanted me to play twentieth century music. Then a friend of mine had a teacher who I went to and she was much nicer. But I still didn't like that. It didn't occur to me to say, I want to play so and so. Because I was a child, I was used to just being told what to do. The scores for the Johann Sebastian* Bach's Double Violin Concerto in D minor, BWV 1043 *(middle movement) and the third movement from* Ludwig van Beethoven's Piano Sonata No. 8 in C minor, Op. 13 Pathetique *were given to us as part of the school music course, not to perform but to study. But I tried playing them and had fun. Then I bought the full scores of each. Then more Beethoven scores, and tried to play them. Also transcribed the Bach and tried to play it. Then some Mozart scores and got sucked into that Rondo in A minor. For a while I thought that the Bach violin concerto reconstructed from the harpsichord transcription BWV 1052R was even better but I changed my mind back. I've never heard either of those played to my satisfaction'.*

Thomas C. Südhof (born 22.12.1955)—is a German-American biochemist and neuroscientist, currently Avram Goldstein Professor in the School of Medicine at Stanford University (Professor, Department of Molecular & Cellular Physiology and of Neurosurgery; Professor (by courtesy), Department of Neurology & Neurological Sciences and of Psychiatry & Behavioral Science). In 2013 he was awarded the Nobel Prize in Physiology or Medicine, shared with James Rothman and Randy Schekman. In addition to being a distinguished scientist, Prof. Südhof is also a bassoonist, and emphasizes the importance of music for his scientific career. In particular, he highlights 'the value of disciplined study, or repetitive learning, for creativity. You cannot be creative on a bassoon if you don't know it inside out, and you cannot be creative in science if you don't have a deep knowledge of the details' (Romine 2013). Prof. Südhof called his bassoon teacher—Herbert Tauscher—as his 'most influential teacher'. When he was asked the question 'You can have dinner tonight with a famous person—who would it be?', Prof. Südhof replied, 'Wolfgang Amadeus Mozart, so that I could try and find out if his creativity was conscious or inherent' (The Lancet 2010) (Fig. 6.7).

Saul Perlmutter (born 22.09.1959)—is an American astrophysicist, a professor of physics at UC Berkeley, where he holds the Franklin W. and Karen Weber Dabby Chair, head of the International Supernova Cosmology Project at the Lawrence Berkeley National Laboratory, and a senior scientist at Lawrence Berkeley National Laboratory. He, along with Adam Riess and Brian P. Schmidt, was awarded the Nobel Prize in Physics in 2011. Perlmutter is also a violinist and teaches at UC Berkeley a course called 'Physics & Music'.

Bob Coecke (born 23.07.1968)—is a Belgian theoretical physicist and logician who was Professor of Quantum Foundations, Logics and Structures at the University of Oxford (2011–2021). Currently he is Chief Scientist at Quantinuum and Distinguished Visiting Research Chair at the Perimeter Institute for Theoretical Physics in Canada. Bob is known for pioneering Categorical quantum mechanics (2004), Quantum Picturalism (2017, 2023), ZX-calculus (2008, 2011), DisCoCat (Categorical Compositional Distributional, 2010), Quantum natural language processing (QNLP, 2016, 2020)

Fig. 6.7 Thomas C. Südhof with bassoon (Photo from Thomas C. Südhof's private archive)

and DisCoCirc (2019). In addition to this, he is also a musician and composer who plays guitar. Bob and his music group 'Black Tish' have been retrospectively called pioneers of industrial music. He was also one of the first to use quantum computers and quantum hardware in music composition and research (e.g. the 'Quanthoven' article—*A Quantum Natural Language Processing Approach to Musical Intelligence* 2021).

The list of important scientists interested in music and art can be continued.

References

Abramsky, S., and B. Coecke. 2004. A Categorical Semantics of Quantum Protocols. *Proceedings of the 19th IEEE conference on Logic in Computer Science (LiCS'04)*. IEEE. arXiv:quant-ph/0402130.

Bach, J.S. 1730. *Concerto for 2 Violins in D minor*, BWV 1043.

Bach, J.S. 1734. *Violin Concerto in D minor*, BWV 1052R.

Beethoven, L. van. 1799. *Piano Sonata No. 8 in C minor*, Op. 13, *Pathétique*.

Beethoven, L. van. 1806. *Violin Concerto in D major*, Op. 61.

Beethoven, L. van. 1807. *Piano Sonata No. 23 in F minor*, Op. 57, *Appassionata*. Vienna: Bureau des Arts et d'Industrie.

Beethoven, L. van. 1809. *Piano Concerto No. 5 in E-flat major*, Op. 73, *Emperor*.

Beethoven, L. van. 1822a. *Piano Sonata No. 31 in A Flat major*, Op. 110.

Beethoven, L. van. 1822b. *Piano Sonata No. 32 in C minor*, Op. 111.

Blum, B. 2002. *Heisenberg and Music*. Heisenberg Family Archive. https://web.archive.org/web/20180831103506/http://heisenbergfamily.org/hbgmusik.htm (Accessed 19 March 2023).

Blum, B. 2020. Musik und Philosophie—Quellen der Kreativität bei Werner Heisenberg. In *Quanten* 8, ed. K. Kleinknecht. Stuttgart: S. Hirzel Verlag.

Carson, C. 2014. *Heisenberg in the Atomic Age: Science and the Public Sphere*. New York: Cambridge University Press.

Coecke, B., and R. Duncan. 2008. Interacting Quantum Observables. *Automata, Languages and Programming, Lecture Notes in Computer Science*, vol. 5126. Springer.

Coecke, B., M. Sadrzadeh, and S. Clark. 2010. Mathematical Foundations for a Compositional Distributional Model of Meaning. arXiv:1003.4394.

Coecke, B., and R. Duncan. 2011. Interacting Quantum Observables: Categorical Algebra and Diagrammatics. *New Journal of Physics* 13 (4): 043016.

Coecke, B., and A. Kissinger. 2017. *Picturing Quantum Processes: A First Course in Quantum Theory and Diagrammatic Reasoning*. Cambridge: Cambridge University Press.

Coecke, B. 2019. The Mathematics of Text Structure. arXiv:1904.03478.

Coecke, B., G. de Felice, K. Meichanetzidis, and A. Toumi. 2020. Foundations for Near-Term Quantum Natural Language Processing. arXiv:2012.03755.

Coecke, B., and S. Gogioso. 2023. *Quantum in Pictures*. Quantinuum.

Deutsch, D. 1997. *The Fabric of Reality*. Penguin Books.

Deutsch, D. 2011. *The Beginning of Infinity: Explanations that Transform the World*. Penguin Books.

Gottfried, K., and J.D. Jackson. 2003. Mozart and Quantum Mechanics: An Appreciation of Victor Weisskopf. *Physics Today*. 56 (2): 43–47.

Heisenberg, W., and E. Heisenberg. 2016. *My Dear Li: Correspondence, 1937–1946*, ed. A.M. Hirsch-Heisenberg, trans. I. Heisenberg. New Haven & London: Yale University Press.

Heisenberg, W. 2019. *Reality and Its Order*, ed. K. Kleinknecht, trans. M.B. Rumscheidt, N. Lukens and I. Heisenberg, Introduction by H. Rechenberg, Commentary by E.P. Fischer. Springer Nature.

Josephson, B.D., and T.L. Carpenter. 1991. Musical Minds. *New Scientist*.

Josephson, B.D., and T.L. Carpenter. 1994. Music and Mind—A Theory of Aesthetic Dynamics: On Self-Organization. *Springer series in Synergetics*, Vol. 61, 280–287. Heidelberg: Springer.

Josephson, B.D., and T.L. Carpenter. 1996. What Can Music Tell Us About the Nature of the Mind? A Platonic Model. In *Toward a Science of Consciousness*, ed. Stuart R. Hameroff, Alfred W. Kaszniak, and Alwyn C. Scott. Cambridge, MA: MIT Press.

Kleinknecht, K. 2019. *Einstein and Heisenberg: The Controversy Over Quantum Physics*. Springer Nature.

Kleinknecht, K., ed. 2020. *Quanten 8*. Stuttgart: S. Hirzel Verlag.

Kozhamkulov, T. 1985. *Obobshchennye stohasticheskie uravneniya v teorii polya* (Generalized Stachastic Equations in Field Theory). Kiev: Academy of Sciences of the Ukrainian SSR. Institute of Theoretical Physics.

Kozhamkulov, T. 1993. *Stohasticheskoe kvantovanie v teorii polya* (Stochastic Quantization in Field Theory). Monograph. Almaty: Ğylym.

Longair, M.S., and J.R. Waldram. 2009. Sir Alfred Brian Pippard. 7 September 1920–21 September 2008. In *Biographical Memoirs of Fellows of the Royal Society*, ed. T.W. Meade, vol. 55, 201–220.

Migdal, A.A., & Kozhamkulov, T.A. (1984) Teoriya vozmushchenij v stokhasticheskom kalibrovochnom pole. Reshetochnyj variant (Perturbation Theory in Stochastic Gauge Field Lattice Version). *Yadernaya Fizika* (Nuclear Physics) 39 (6): 1596–1605.

Miranda, E.R., R. Yeung, A. Pearson, K. Meichanetzidis, and B. Coecke. 2021. A Quantum Natural Language Processing Approach to Musical Intelligence. arXiv: 2111.06741.

Mozart, W.A. 1787. *Rondo in A minor*, KV 511. Vienna: Hoffmeister.

National Academy of Sciences of the Republic of Kazakhstan. 2022. *Enciklopedicheskij spravochnik* (Encyclopedic Reference Book). Almaty: NAS RK.

Nobel Foundation. 2019. *The symphony of science*. https://www.nobelprize.org/symphony-of-science/ (Accessed 4 January 2024).

Pesic, P. 2014. *Music and the Making of Modern Science*. Cambridge, MA: MIT Press.

Politzer, H.D. 1973. Reliable Perturbative Results for Strong Interactions? *Physical Review Letters* 30 (26): 1346–1349.

Rechenberg, H. 2019. Introduction. In *Reality and Its Order*, W. Heisenberg, trans M.B. Rumscheidt, N. Lukens, and I. Heisenberg. Springer Nature.

Romine, R.D. 2013. From Bassoonist to Nobel Laureate: An Interview with Thomas Südhof. *The Double Reed* vol. 36, no. 4.

The Lancet. 2010. Tom Südhof 376 (9739): 409. https://doi.org/10.1016/S0140-6736(10)61210-X.

Waldram, J.R. 2008. *Professor Sir Brian Pippard (1920–2008)*. https://www.cam.ac.uk/news/professor-sir-brian-pippard-1920-2008 (Accessed 15 October 2023).

Wilczek, F. 2021. *Fundamentals: Ten Keys to Reality*. Penguin Books.

Zeng, W., and B. Coecke. 2016. Quantum Algorithms for Compositional Natural Language Processing. *Electronic Proceedings in Theoretical Computer Science* 221: 67–75.

7

Origins and Development of Classical Music in Faces

One of the integral parts of human history is the history of music, and the intellectual spirit of society largely depends on its perception. Entire fields within musicology are devoted to researching the history of music. Most of the works are encyclopedic in nature and, as a rule, incredibly voluminous. Studying them may seem like a daunting task. Sometimes very specific issues are studied in detail.

This work (Abdyssagin 2021, pp. 6–24) fills the need for a brief and clear presentation of the history of the development of Western European music from the Middle Ages to the beginning of the twentieth century. There is an opinion that the whole history of mankind is based on biographies of great people (the 'great men theory'). This chapter serves as an introduction dedicated to an extensive topic of the development of new music. By following some milestones in the history of music we will be able to open doors to an unknown world filled with wonders and delights of contemporary music.

At the age of lightning-fast changes and the highest concentration of information—when the flow of information pouring on us is so diversified that we are literally overwhelmed—it is important to have some reference points that are not subject to change. Some non specialists may perceive the history of musical art and its evolution as something incomprehensibly complex and vast. This may be partially true. However, the aim of this review is to give a clear idea, a transparent picture of the beauty expressed through the development of classical music. The reader can follow the impulse that helped the entire 'tree of music' grow and see the directions of its development, its ramifications and its current state.

R.-B. Abdyssagin, *Quantum Mechanics and Avant-Garde Music*, https://doi.org/10.1007/978-3-031-63161-0_7

For many, this chapter can become an initial entrance ticket to the mystery of the world of composition. In our time it is important for composers to realise their importance, because they are the carriers of a thousand-year history of the development of the greatest of the arts. Nothing goes unnoticed. In the same way music and its history have been intertwined with thousands of events, lives and discoveries leading us to the future. Countless nodes, pivots and 'beacons' of harmony have been created, some of which later invariably dissolved and reappeared to create the art of the new time.

Contemporary music is not pop music. Contemporary/avant-garde music is precisely the evolutionary development of classical music that we all know. To understand its formation, it is first necessary to understand the path that the art of music has taken.

It is known that, like any other art and even more than others, music—which is the most sublime and abstract of all the arts—is the first to catch even the subtlest changes in the development of human consciousness. In its development music reflects key evolutionary moments of the entire human civilisation. In order to predict the future and understand the present trajectories, it is necessary to look into the past. The various biographical details of certain great composers or music eras are not aimed to be addressed in detail. The concentration will be in the most characteristic qualities of great creators and different stories from their lives, so that the light can be shed to their mentality and level of thinking.

Humankind belongs to the species called Homo Sapiens, but it also known that thousands of years ago Homo Sapiens became 'Homo Musicales' or 'Homo Musicus' (a term that appears in Kirnarskaya 2009). Music has always been an integral part of our society and human life. It was used both for the purposes of communication and self-expression. Since prehistoric times music has been one of the highest manifestations of the human mind. Broadly speaking, everything that we possess today and all the achievements that we have at our disposal are the products and phenomena of our civilisation that would not have appeared if we did not have the unique quality of intellect and reasoning.

The word 'classical' has its own definitions. Of course, there may be different understandings, but not every great composer can be called a 'classical' composer because the era of classicism has certain specific and precise 'laws' according to which it develops, not only in music, but also in drama, literature and visual arts. What is usually referred to as 'classical' at the time of its appearance was the most advanced contemporary/avant-garde music. That is many great composers of their time were real avant-garde creators and pioneers in their world.

Since ancient times rhythm has been the fundamental principle of music and percussions were one of the first musical instruments. But along with this the human voice played a huge role, as all other instruments were invented 'in the image and likeness' of the human voice.

There are several classifications of the music history. This chapter will adhere to the main periods. The longest is the era of the Middle Ages, followed by the Renaissance, Baroque, Classicism, Romanticism and the beginning of the twentieth century.

The Middle Ages spans between years ~ 500–~1500. The precise starting and ending dates are still debated among historians. According to historians, the Middle Ages began in 476 when the last Emperor of the Western Roman Empire was deposed, but when exactly it ended still causes many arguments. Some argue that it ended after the capture of Constantinople by the forces of the Ottoman Empire in 1453. Others suggest the first voyage of Christopher Columbus to the Americas in 1492, or the beginning of Protestant Reformation in 1517 when Martin Luther published his *Ninety-five Theses or Disputation on the Power and Efficacy of Indulgences*, and a number of other events. Some even suggest that the end of the Middle Ages was around the beginning of the seventeenth century, when in 1600 the great Giordano Bruno was burned alive at the stake in Rome in Campo de' Fiori. In general, it is possible to say that the Middle Ages ended in the fourteenth century, and the Renaissance emerged and flourished in the fifteenth and sixteenth centuries.

The history of Western European Music began in the Middle Ages. In particular, during the papacy (590–604) of St. Gregory I (~540–604) monophonic liturgical music was collected and codified, and later became known as Gregorian Chant. Charlemagne (~748–814) supported and aggressively imposed its use throughout the Carolingian Empire, and the synthesis of Gallican Rites and Gregorian Chant occurred during the eighth and ninth centuries, thus around the tenth century Gregorian Chant (in its evolved form) was already widely used.

Many of the principles in music that were established during the Middle Ages are still valid to this day. Since that time the pitch and the rhythm have become two fundamental, system-forming coordinates of musical space and time. And only at the beginning of the twentieth century, when sonoristics and new techniques of playing instruments appeared, did the timbre-texture emerged as one of the fundamental coordinates. However, there was a unique attitude toward timbre to a certain extent even before that.

Fig. 7.1 Guido d'Arezzo (*Source* Wikimedia Commons)

Fig. 7.2 Hildegard von Bingen with nuns (*Source* Wikimedia Commons)

During the Middle Ages Guido d'Arezzo (~991/992–~1050) created the first prototype of a modern musical notation system. It is believed that Guido (Fig. 7.1), showing the singers how to sing, held four fingers in front of him and pointed at them to show how approximately the melodic lines should develop. Thus the first prototype of modern notation was based on four lines. Among German composers the great Hildegard von Bingen (~1098–1179) should also be mentioned (Fig. 7.2).

In the Middle Ages the art of polyphony was established as the basis of the entire existence of harmony and architectonical structures of music. At that time the major part of music was inseparable from the church. The great genres of polyphony such as motet and organum, as well as isorhythmic technique appeared. The isorhythmic technique is still used today in certain ways. Some echoes can be seen in the *Four Rhythmic Etudes* (1950) by the great French composer of the twentieth century Olivier Messiaen, although this technique originated in the Middle Ages. In simple terms, the essence of the isorhythmic technique is when the rhythmic formula itself, one postulate, does not change rhythmically and is repeated many times. When it is repeated the rhythm remains unchanged but the 'melody' changes constantly.

Then there was the famous school of Notre Dame, to which the great composers Léonin (~1150–~1210) and Pérotin (~1160–~1230) belonged. In addition to the development of polyphony and the first techniques of counterpoint, the Notre Dame school and especially the compositional methods of Léonin and Pérotin were of exceptional significance because

Fig. 7.3 A bell player, considered to be Perotinus Magnus (*Source* Wikimedia Commons)

Fig. 7.4 Albertus Magnus (Fresco by Tommaso da Modena. *Source* Wikimedia Commons)

during that period for the first time rhythm/durations gained as much importance as melody/pitch. While Léonin was famous for writing masterly writing two-voice music, Pérotin (Fig. 7.3) was recognised for his works using three or more voiced polyphony.

At that time—as believed by some specialists—not only the composers, but in general all artists were treated like artisans. Later, after Cimabue, Giotto in visual arts, Dante and Boccaccio in literature, Léonin and Pérotin in music, a composer started to be treated as an individual capable of creating something unique. Pérotin was the first to be called a great composer, his nickname was Perotinus Magnus, which means Pérotin the Great. Thanks to them, later society came to realise that creativity is a phenomenon that is unique and unrepeatable. It can be said that the Western musical tradition (in the way we know it) began from Pérotin. The naming of Perotinus Magnus has some connection with the naming of Albertus Magnus who was one of the most influential and significant medieval philosophers and theologians (Fig. 7.4).

The work of the composers of the Ars Antiqua (ancient art) era can be compared with the approach of the avant-garde for many of the principles of the musical avant-garde as well as many methods were introduced at that time. Ars antiqua was followed by the Ars Nova (new art).

The great French composer Philippe de Vitry (1291–1361) wrote the eponymous treatise *Ars Nova* (~1322, Fig. 7.5). At the same time the new concept of 'mensural notation' was introduced. This is the first prototype of modern rhythm notation, when the rhythm could already be indicated separately and clearly, regardless of the pitch line. Guillaume de Machaut

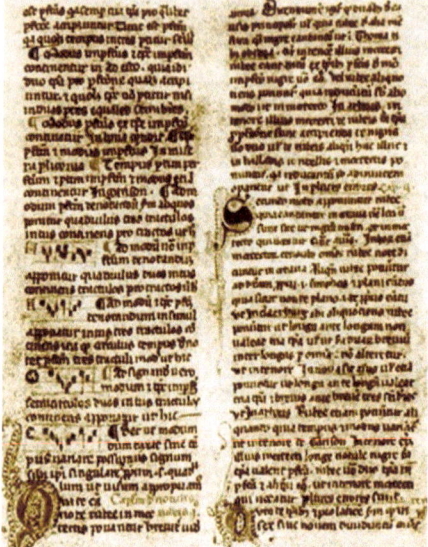

Fig. 7.5 Page from Philippe de Vitry's treatise (*Ars Nova* Musicae, Public domain. *Source* Wikimedia Commons)

Fig. 7.6 *Kyrie* from Guillaume de Machaut (*Messe de Nostre Dame*, Public domain. *Source* Wikimedia Commons)

(~1300–1377), the author of the first mass—*Messe de Nostre Dame* (Fig. 7.6) and Francesco Landini (c. 1325/1335–1397) were the bright representatives of that epoch. It was the gracefulness, elegance of Landini's musical themes, the peculiarities of handling polyphonic lines, the structuring of certain musical elements that largely determined the development of Italian art for the next century.

At that time many composers, painters and writers worked at the court. For example, the outstanding artists—the Limbourg brothers—served at the court of the Duke of Berry (Fig. 7.8).

Folk art also developed. Its representatives were troubadours in southern France. Guillaume IX of Aquitaine (1071–1126) is considered to be the first troubadour. His son Guillaume X had a daughter—Eleanor of Aquitaine (~1122–1204), who was the Queen of France, and later the Queen of England. And the son of Eleanor of Aquitaine, who was the great-grandson of Guillaume IX, was the English King Richard I the Lionheart (1157–1199), a famous trouvère as well. If there were troubadours in southern France, then in the north during the late Middle Ages there were trouvères. One of them was Thibault IV of Champagne (1201–1253).

In German lands they were called minnesingers. Their art was called 'minnesang' (love song). The most famous was Walther von der Vogelweide

Fig. 7.7 F. Landini (From *Codex Squarcialupi. Source* Wikimedia Commons)

Fig. 7.8 Tres Riches Heures du duc de Berry, June (By Limbourg brothers, Musée Condé. Public domain. *Source* Wikimedia Commons)

Fig. 7.9 W. von der Vogelweide (From *Codex Manesse. Source* Wikimedia Commons)

(~1170–~ 1230) (Fig. 7.9) and the last of the Minnesingers was Oswald von Wolkenstein (~1376/1377–1445).

Touching the overall context of that late medieval period, philosophy and theology played significant roles in the intellectual and spiritual life of society in Europe. Ideas and works of Anselm of Canterbury (~1033/1034–1109), Albertus Magnus (~1200–1280), Thomas Aquinas (1225–1274), Roger Bacon (~1219/1220–~1292), Bonaventure (1221–1274), John Duns Scotus (~1265/1266–1308), William of Ockham (~1285–1347, celebrated for the principle called *Ockham's razor*) and many others.

The Renaissance era followed the Middle Ages. During the Renaissance, one of the most prominent composers was Giovanni Pierluigi da Palestrina (1525–1594), who is considered to be among the founders of the historic Italian composition school (Fig. 7.10)

It is believed that it was during that era that the genre of opera was born. The first interludes, the forerunners of the opera, were commissioned by Cosimo I de Medici and Lucrezia Borgia, and were dedicated to them. It was in the period from the Renaissance to the Baroque and later that the true measure of the wealth, influence and power of this or that duke or ruler was determined by which composer would write or dedicate operas to them.

Fig. 7.10 G.P. da Palestrina (*Source* Wikimedia Commons)

Fig. 7.11 C.Monteverdi (Portrait by Bernardo Strozzi. Tiroler Landesmuseum Ferdinandeum. *Source* Wikimedia Commons)

For example, the opera that predetermined the further development of the genre— *L'Orfeo* (1607) by Claudio Monteverdi (1567–1643) (Fig. 7.11) — was dedicated to the Duke of Mantua Francesco Gonzaga. Similarly the great Antonio Vivaldi (Fig. 7.12) dedicated his works to the rulers of his time, Christoph Willibald Gluck—to the Emperor of Austria, and the great Franz Joseph Haydn, who served the princes of Esterhazy for more than thirty years, composed a quartet called *The Emperor*, as well as a cycle of string quartets Op. 50 dedicated to the King of Prussia Friedrich Wilhelm II. The most ambitious concert of the era of classicism—*Piano Concerto No. 5* in E-flat major, Op. 73 by Ludwig van Beethoven—is also known as the *Emperor Concerto*.

The Baroque era was marked by several great Italian composers. We can name many of them but one of the most prominent was Antonio Vivaldi (1678–1741), who is known to this day, especially for his concerts for string instruments as well as the cycle Le quattro stagioni (The Four Seasons). That was the time when the main distinctive features of the Venetian school of composition were born—a unique attitude to space. According to the current Italian composers and intellectuals Alessandro Solbiati and Gabriele Manca, many features of how Vivaldi treated and structured musical space can be traced in the works of the great Venetian composers of the twentieth century—Luigi Nono and Bruno Maderna.

1685 was a unique year, *Annus Mirabilis* in the history of music, for in this year three great representatives of the Baroque era were born—Johann

Fig. 7.12 A. Vivaldi (Museo internazionale e biblioteca della musica, Bologna. *Source* Wikimedia Commons)

Fig. 7.13 G.F. Handel (Portrait by Philippe Mercier. *Source* Wikimedia Commons)

Fig. 7.14 D. Scarlatti (Portrait by Domingo Antonio Velasco. *Source* Wikimedia Commons)

Sebastian Bach (1685–1750), Georg Friedrich Handel (1685–1759), and Domenico Scarlatti (1685–1757).

Handel (Fig. 7.13) gained recognition at the age of 19 when his first opera *Almira, Königin von Castilien* was staged. Later he traveled to Italy and visited the Medici. At the court of Cardinal Ottoboni he met Domenico Scarlatti (Fig. 7.14). They arranged a clavier competition and as a result Scarlatti was recognised as the best harpsichord performer while Handel was recognised as the best organist. Later Handel moved to England.

When Franz Joseph Haydn (Fig. 7.15) visited London, he learned that the King had bought Handel's clavier as a great legacy of the outstanding composer (Dies 1810, p. 112). An interesting fact—when Handel was born, his father was 63 years old! And after the birth of Handel his father had 3 more children.

Ludwig van Beethoven enormously admired Handel and even said, 'Handel is the greatest composer that ever lived… I would uncover my head and kneel down on his tomb'. When I was visiting London in May 2014 and performing my work *Kazakh rhapsody* for piano and symphony orchestra (2014) at the Royal Festival Hall, I also especially visited the famous Westminster Abbey. Apart from English monarchs, Westminster Abbey is a place where Isaac Newton—along with many other outstanding physicists—and the great Georg Friedrich Handel are buried. I was happy to be able to kneel down on the tomb of Handel.

Johann Sebastian Bach had a completely different fate. It is believed that he never traveled outside Germany all his life, but when in 1977 The Voyager Golden Records was sent to space with Voyager-1 and Voyager-2

Fig. 7.15 F.J. Haydn (Portrait by Thomas Hardy, Royal College of Music, Museum of Instruments, London)

Fig. 7.16 W.A. Mozart (Portrait by Joseph Lange, Mozarts Geburtshaus, Salzburg)

Fig. 7.17 L. van Beethoven (Portrait by Joseph Karl Stieler, Beethoven-Haus, Bonn)

containing some information about humanity, they carried selected works of Johann Sebastian Bach performed by Glenn Gould, as well as music by other great composers. Bach was an amazingly productive creator who wrote *Das Wohltemperierte Klavier* I, II (BWV 846–893), *Die Kunst der Fuge* (BWV 1080), *Matthäus-Passion* (BWV 244), *Johannes-Passion* (BWV 245) a huge number of cantatas and pieces for many types of claviers as well as pipe organ. His children Carl Philipp Emanuel Bach, Johann Christian Bach, Wilhelm Friedemann Bach became prominent composers. In particular, the great Joseph Haydn studied using the book of C.P.E. Bach. It is known that Johann Christian Bach supported and highly appreciated the young Wolfgang Amadeus Mozart. It is believed that Bach had 20 children. Unfortunately, at that time there was a high mortality rate among children and only a few of them survived. Robert Schumann, the great composer, once said: 'Music owes as much to Bach as religion to its founder'.

Classical music was born after the emergence of Isaac Newton's classical mechanics and law of universal gravitation. The era of classicism brought the understanding that tonality plays as important a role in music as gravity does in a physical space. The transformation of the musical foundations took place gradually, starting with the works of Bach (the Baroque era), right up to the great classical composers (classicism). A huge leap took place, the all-embracing polyphony and its absolutism moved on to the triumph of classical harmony and tonality in music.

The era of classicism cannot be imagined without the great Wolfgang Amadeus Mozart (1756–1791). He lived for only 35 years but composed over 600 works. This is probably the greatest legacy of our entire civilisation. Great Mozart still remains the measure of the genius of mankind. The height of the development of the human mind and our civilisation can be determined and judged by listening to the creations of Mozart (Fig. 7.16).

Many great people are only acknowledged after their deaths. In order to understand and truly appreciate the greatness of one person, the one who evaluates must be at approximately the same level. Johann Sebastian Bach was almost consigned to oblivion after his death. Although he was considered to be a great master during his lifetime and a respected person, his works were believed to be too difficult to comprehend. Seventy-nine years after the death of J.S. Bach, Felix Mendelssohn revived Bach's music. Only the greatest of composers—F.J. Haydn, W.A. Mozart and L. van Beethoven understood and comprehended the genius of Bach since they themselves were geniuses. As Beethoven (Fig. 7.17) said, 'Nicht Bach, sondern Meer sollte er heißen, wegen seines unendlichen, unerschöpflichen Reichtums an Tonkombinationen und Harmonien. Bach ist der Urvater der Harmonie' (Beethoven-Haus Bonn, no date). Beethoven's phrase can be translated into English as 'Not Bach (brook), but ocean should be his name because of his infinite, inexhaustible wealth of tonal combinations and harmonies. Bach is the forefather of harmony'.

Many are surprised by how in the novel *The Great Gatsby* (1925) by Francis Scott Fitzgerald (1896–1940) Gatsby himself had enormous success in society. And when Gatsby died, only his father, the priest and the man on whose behalf the novel was written, were at his funeral. When Wolfgang Amadeus Mozart died he had no grave. According to the legend, he was so poor that his family did not have enough money to bury him and he was thrown into a common grave (however, there is an opinion that burying in a common grave was a custom of that time for ordinary people). The location of his grave is still rather approximate. Such are the vicissitudes of fate when a person whose genius determined the level of development of the entire civilisation does not have his own grave, which is a great tragedy (Fig. 7.18).

Both Haydn and Mozart supported Ludwig van Beethoven (1770–1827). Many musical laws and principles that are important to this day are associated with the name of Beethoven. In his work, the principles of the sonata, the form of sonata, and the very idea of dialectics in music reached their apogee.

L. van Beethoven raised the composer's art and personality to unprecedented heights. The premiere of his Symphony No. 9 d-moll Op. 125 received a standing ovation six times. Moreover, at that time Beethoven was already completely deaf. He could hear neither the performance nor the enthusiastic jubilation of the audience. He did not look at the audience being afraid of a misunderstanding. Contralto Caroline Unger turned Beethoven to

Fig. 7.18 First page of the manuscript of *Requiem in D minor, KV 626*, the last and one of the most famous works by Wolfgang Amadeus Mozart. Public domain. Österreichische Nationalbibliothek, Wien (*Source* IMSLP, Petrucci Music Library)

face the audience so that he could see a standing ovation honoring his music and genius.

When I was practicing Beethoven's *Sonata No. 23* f-moll Op. 57 *Appassionata*, I thought, 'to what despair Beethoven had to be driven, and what inner core he must have had in order to endure all this, and write such a sonata and such a texture for the piano, and even the hand movements'. In part, this sonata is an example of the birth of a new pianism, a completely innovative interaction with the instrument for that time.

Beethoven certainly has a great piano legacy. All 32 of his piano sonatas are absolutely brilliant masterpieces. Beethoven's last Piano Sonatas, Nos. 30, 31 and 32 represent the pinnacle of his piano thought. Sonata No. 30 E-Dur Op. 109 immerses listener in a new world of harmony and form. Sonata No. 31 As-Dur Op. 110 touches on the deepest threads of human psychology and the inner world, raises global issues of time and existence. Klagender Gesang (Fig. 7.19) in particular has the most piercing impact. The spirit of a man, the creator, rises above the entire universe and casts a deep gaze on all earthly processes. And the last Sonata No. 32 c-moll Op. 111 creates a space of new

Fig. 7.19 Excerpt from the manuscript of L. van Beethoven *Sonata No. 31 As-Dur* op. 110 (*Klagender Gesang* section. Public domain. Staatsbibliothek zu Berlin. *Source* IMSLP, Petrucci Music Library)

music (Fig. 7.20). Beethoven anticipates jazz rhythms more than 100 years before the appearance of jazz!

Many of Beethoven's works were dedicated to the noble people of his time. In particular, *Symphony No. 9* mentioned above was dedicated to the King of Prussia Friedrich Wilhelm III, Sonata for Violin and Piano No. 6 A-Dur Op. 30 was dedicated to the Russian Emperor Alexander I. Count Waldstein, Princes K.Lichnowski, J.F.M.Lobkowitz, A.Razumovsky were influential, prominent and the richest people of their time. But now they are remembered and their name remained in history largely due to the fact that Beethoven dedicated his works to them!

Moving on to the next eras the great Carl Czerny (1791–1857) should be noted. While Muzio Clementi (1752–1832) was called 'Padre del Pianoforte—'the father of piano—Carl Czerny was the one who created the modern pianism. In the nineteenth century there was hardly a single famous pianist who did not study with him, was not familiar with the 'school of Czerny'. Franz Liszt was one of his students and Czerny himself was a student of Beethoven. Czerny had a phenomenal memory. When he was 14 years old, Beethoven wrote a recommendation letter thus completing his studies. At the age of 15 he began teaching. It is believed that for several decades, until his death, Czerny gave up to 12 lessons a day, and wrote music at night.

C. Cherny created 861 opuses. One opus can contain either one or several dozen pieces. Czerny has opuses that contain over a hundred of pieces. A huge

Fig. 7.20 Excerpt from the manuscript of L. van Beethoven *Sonata No. 32 c-moll* (Section where revolutionary complex rhythms predict 'jazz'. Public domain. Staatsbibliothek zu Berlin. *Source* IMSLP, Petrucci Music Library)

number of piano studies, in particular, Opus 740 (50 Etudes) are the foundation of all modern piano playing tradition. I would even say that Carl Czerny is among the most hard working geniuses in the history of music. By the way, we, graduates of Master in II Livello (doctoral studies) of the Conservatorio Statale di Music 'Cesare Pollini' di Padova (2019), we have a direct connection to the noble pianistic pedigree. So: Rakhat-Bi Abdyssagin → Konstantin Bogino → Vera Gornostayeva → Heinrich Neuhaus → Alexander Siloti → Franz Liszt → Carl Czerny → Ludwig van Beethoven → Wolfgang Amadeus Mozart → Johann Christian Bach → Johann Sebastian Bach. This is clear evidence that the school of C.Cherny, his legacy lives 'in hands' of almost all pianists on the planet. And even from the heights of today we can talk about a continuous line of development of pianism from the great creators of those times to the masters of the present.

It is believed that Czerny usually came to the concert hall and asked the audience to tell him any number from 1 to 32, which is the amount of Beethoven's sonatas (32). Czerny knew them all by heart and was ready to play them at any moment. He also edited works by Johann Sebastian Bach, Domenico Scarlatti, Wolfgang Amadeus Mozart and many others. As many teachers say, without Czerny's works they would not be able to imagine their teaching activities and practice. C. Czerny was the greatest teacher (in a wide sense of the word) in the history of music.

In the era of classicism the rules of dialectics such as 'the unity and conflict of opposites', 'the transformation of quantity into quality', and 'the negation of the negation' were reflected in music, particularly in the principle of sonata (sonata is not a mere form but a major philosophical, metaphysical and ontological principle in music). The principle of 'negation of negation' could be one of the possible symbols of development within the dramaturgical dialectical form of a piece of music.

In general, as one of the possible interpretations, there are three main creative methods: avant-garde, classicism, and romanticism. Avant-garde is tracing of new directions and when everything old is made obsolete. People dwell into the unknown. The era of classicism usually emerges after the avant-garde, marking the ideal, the highest peak in the development of a particular idea, when everything revolves around something unified. And after the era of classicism, the era of romanticism comes, when the old forms, old principles are no longer able to reflect the new course of time, new ideas and the new understanding of the structure of the universe. Then the old forms collapse and the search for new ones begins. After that again comes avant-garde, classicism, romanticism, and this spiral movement is among the possible models of the development of music in time.

The era of romanticism is known for such composers as Franz Peter Schubert (1797–1828), Hector Berlioz (1803–1869), Robert Schumann (1810–1856), Frédéric François Chopin (1810–1849), and many others. Entire treatises could be devoted to each of them. Franz Peter Schubert deeply admired Beethoven, and was even buried next to him. Schubert lived for only 31 years but during this short time made a huge contribution to the development of music.

Frédéric Chopin was a great composer, and probably there is no person who would not know his name. Chopin was a true poet of piano. And without Franz Liszt (1811–1886), even the piano itself would not be what it is today. There is a view that the range of a piano was expanded thanks to the creative influence of Liszt. There were many types of claviers. These include the virginal, spinet, harpsichord, clavichord and others. And, in turn of the seventeenth and eighteenth centuries, Bartolomeo Cristofori (1655–1731), originally from the city of Padua in Italy, invented a new type of instrument—the first prototype of a modern piano. The invention of piano became the greatest achievement not only of Cristofori, but also one of the hallmarks and turning points in the history of keyboard music.

Fig. 7.21 Rakhat-Bi Abdyssagin plays P.I. Tchaikovsky's personal Grand Piano (P.I. Tchaikovsky's House-Museum, Klin, Russia, 20.09.2018. Photo: Rakhat-Bi Abdyssagin's private archive)

The great Felix Mendelssohn (1809–1847) performed Johann Sebastian Bach's *St Matthew Passion* in 1829 (at the age of 20). After this concert, people started to consider Bach one of the greatest composers ever.

In Russia, speaking about its art, it is impossible not to mention Pyotr Ilyich Tchaikovsky (1840–1893) and Nikolai Andreyevich Rimsky-Korsakov (1844–1908). One was a representative of Moscow, the other of Saint Petersburg composition schools.

I am exceptionally honoured that on 20th September 2018 I had the privilege of performing piano recital on P.I. Tchaikovsky's personal Grand Piano (Fig. 7.21) in his House-Museum in Klin, Russia (I was invited by the Association of Tchaikovsky Competition Stars, which I am a member of). For me, playing on the Grand Piano that contains a reflection of Tchaikovsky's soul is like communicating with Tchaikovsky himself. This Grand Piano has a truly amazing and unique aura and in many ways is a reflection of the essence and worldview of Tchaikovsky himself. When I started to play it, the instrument almost began to teach and guide me, I felt a deep connection with this Grand Piano, it seemed that it even has its own mind and reason. Just touching the keys of this instrument was like touching something divine.

In France there were Claude Debussy (1862–1918) and Maurice Ravel (1875–1937), great impressionist and symbolist composers.

There was Richard Wagner (1813–1883) in Germany (was son-in-law of Franz Liszt, and was married to Liszt's daughter Cosima). He was the author of operas of unprecedented scale. In his works the tonal system begins to disintegrate and the search for new means of musical expression begins. Tonality in the form in which classical composers knew it plays a completely different role.

Interesting to mention that the revival of the music of Bach, which happened during the Romantic period, and the revival of the music of Vivaldi during the early years of the twentieth century, can be interpreted as one of the reflections of the principle of 'negation of negation'.

Interesting story is that Giuseppe Verdi (1813–1901)—one of the greatest ever opera composers, a true symbol of Italian operatic music—twice failed to get into the Milan Conservatory. First time Verdi was not accepted as a student, and the second time he was rejected as a teacher. And now this conservatory is named after Giuseppe Verdi.

One of the unique features of Italy, a great musical country, is that in almost every town, no matter how small, there is a fantastic opera theatre (which stages wonderful opera productions), and a conservatory, higher education institution in music. And usually a conservatory bears the name of a great composer or musician from this very area. There is a large number of examples of this. This feature is also prominently present in another great musical country—Germany, where brilliant Musikhochschulen (a higher education institution in music) are present everywhere across the country.

There are many amazing and unique processes in the interaction between visual arts and music. For example, Wassily Kandinsky—a painter and one of the pioneers of abstract art in the Western world, was a friend of composer Arnold Schoenberg. Schoenberg himself also loved to paint occasionally. Kandinsky was greatly inspired by music, perceived tones and sounds in colours, and said that 'Color is the keyboard, the eyes are the hammers, the soul is the piano with many strings. The artist is the hand that, by touching this or that key, deliberately causes vibrations of the soul'.

For me, the paintings of Caspar David Friedrich (1774–1840) evoke associations with the music of Romantic composers such as Schubert, Schumann, Liszt and others, while the famous painting *Nighthawks* by the American artist Edward Hopper (1882–1967) evokes associations with the music of American composers such as George Gershwin, Charles Ives, John Cage, George Crumb etc.. And Jackson Pollocks *Blue Poles* (*Number 11, 1952*) largely reflects the direction of sonoristics and a number of other movements in new music.

All contemporary (classical) music, which is sometimes referred to as an avant-garde, is an evolutionary development of classical music. There is a huge number of techniques and systems of new music, be it sonoristics, spectral technique, structuralism, dodecaphony, musique concrète instrumentale, electronic music and many other directions.

Composers of the past created the absolute maximum from the minimum elements; be it a scale, a chord or an arpeggio, they built entire musical palaces from these elements. This is the essence of a composer's art. And now having, let's say, much more developed elements, we can create new universes. If composers of the past had bricks and built 'Versailles', now we have 'nano or even quantum technologies' and we can create whole worlds, new universes.

References

Abdyssagin, R.-B. 2014. *Kazakh Rhapsody* for Piano and Symphony Orchestra.

Abdyssagin, R.-B. 2021. *Kaplya Vechnosti* (A Drop of Eternity). Almaty: Kazak Universiteti.

Bach, J.S. 1722. *Das wohltemperierte Klavier I*, BWV 846–869.

Bach, J.S. 1724. *Johannespassion*, BWV 245.

Bach, J.S. 1727. *Matthäuspassion*, BWV 244.

Bach, J.S. 1740. *Das wohltemperierte Klavier II*, BWV 870–893.

Bach, J.S. 1751. *Die Kunst der Fuge*, BWV 1080. Berlin: C.P.E.Bach.

Beethoven, L. van. 1803. *Violin Sonata No. 6 in A major*, Op. 30 No. 1. Vienna: Bureau des Arts et d'Industrie.

Beethoven, L. van. 1807. *Piano Sonata No. 23 in F minor*, Op. 57, *Appassionata*. Vienna: Bureau des Arts et d'Industrie.

Beethoven, L. van. 1809. *Piano Concerto No. 5 in E-flat major*, Op. 73, *Emperor*.

Beethoven, L. van. 1821. *Piano Sonata No. 30 in E major*, Op. 109. Berlin: Schlesinger.

Beethoven, L. van. 1822a. *Piano Sonata No. 31 in A Flat major*, Op. 110. Vienna: Cappi & Diabelli.

Beethoven, L. van. 1822b. *Piano Sonata No. 32 in C minor*, Op. 111. Berlin: Schlesinger.

Beethoven, L. van. 1824. *Symphony No. 9 in D minor*, Op. 125.

Beethoven-Haus Bonn (no date) *Zitate von Beethoven*. https://internet.beethoven.de/de/ausstellung/beethoven-in-briefmarken/id26.html (Accessed 28 December 2023).

de Machaut, G. 1365. Messe de Nostre Dame.

Dies, A.C. 1810. *Biographische Nachrichten von Joseph Haydn nach mündlichen Erzählungen desselben entworfen und herausgegeben von Albert Christoph Dies.* Wien: Camesinaische Buchhandlung.

Fitzgerald, F.S. 1925. *The Great Gatsby.* New York: Scribner.

Handel, G.F. 1705. *Almira, Königin von Castilien,* Opera.

Kirnarskaya, D. 2009. *The Natural Musician: On Abilities, Giftedness, and Talent,* trans. Mark H. Teeter. Oxford: Oxford University Press.

Messiaen, O. 1950. *Quatre Études de rythme.* Paris: Éditions Durand.

Monteverdi, C. 1607. *L'Orfeo, favola in musica* (opera).

Mozart, W.A. 1791. *Requiem in D minor,* KV 626.

Vivaldi, A. 1720. *Le quattro stagioni.*

8

Origins and Development of Contemporary Music in Faces

In the second half of the twentieth century the general consensus in the scientific world was that physics is completed, almost everything is known (except a few small gaps), and nothing new can be discovered about the world. In a sense, many people believed to have accumulated the completed knowledge about reality. Albert Abraham Michelson—a physicist known for the Michelson-Morley experiment (which determined that light propagates at the same speed in all directions no matter what the motion of the laboratory is, and refuted the aether theories), and a winner of the 1907 Nobel Prize in Physics (the first American to receive a Nobel Prize in the scientific field) —said during the dedication of Ryerson Physical Laboratory in 1894:

> While it is never safe to affirm that the future of Physical Science has no marvels in store even more astonishing than those of the past, it seems probable that most of the grand underlying principles have been firmly established and that further advances are to be sought chiefly in the rigorous application of these principles to all the phenomena which come under our notice. It is here that the science of measurement shows its importance – where quantitative work is more to be desired than qualitative work. (The University of Chicago 1896, p. 159)

In 1878, when young Max Planck studied physics at the Ludwig Maximilian University of Munich, his professor Johann Philipp Gustav von Jolly advised him not to continue in physics because all the fundamental laws of physics had already been discovered. Planck replied that he does not want to make discoveries, but wants to understand the world (Lightman 2005).

R.-B. Abdyssagin, *Quantum Mechanics and Avant-Garde Music*, https://doi.org/10.1007/978-3-031-63161-0_8

As Planck himself recounts what Jolly said him: 'he described physics to me as a highly developed, nearly fully matured science, that [...] will arguably soon take its final stable form. It may yet keep going in one corner or another, scrutinizing or putting in order a jot here and a tittle there, but the system as a whole is secured, and theoretical physics is noticeably approaching its completion' (Wells 2016).

Ironically, Planck—who did not want to make discoveries, and only wished to understand how the world works—made the discovery that shook science and our perception of reality to its very core. Twice ironically, Planck himself, who was a man of very Newtonian views, initially did not want to believe in his own discovery, and was reluctant to accept it. For this Planck was called a 'reluctant revolutionary'.

The following is a brief chronology:

1897 – Joseph John Thomson announces the existence of electron, the first subatomic particle that was discovered. This opened the gates to the microworld.

1900 – Max Planck discovers 'energy quanta' and postulates Planck's constant.

1905 – Albert Einstein (who was 26 years old at that time) published 4 breakthrough papers, including on special relativity and the photoelectric effect.

1911 – Ernest Rutherford introduced the concept of atomic nucleus.

1913 – Niels Bohr (who was 28 years old) formulated his orbital model of a hydrogen atom. Bohr's model was later elaborated by Arnold Sommerfeld.

1914 – James Franck and Gustav Ludwig Hertz conducted an experiment that demonstrated the quantum nature of atom.

1915 – Albert Einstein publishes his General Theory of Relativity (also known as Einstein's theory of gravity, which introduced the curvature of spacetime).

1922 – performance of Stern-Gerlach experiment that showed quantisation of the spatial orientation of angular momentum.

1924 – Louis de Broglie (32 years old) introduced the theory of matter waves.

1924 – the term 'quantum mechanics' appears in the article by Max Born.

1924 – publication of Bose-Einstein statistics, the 'constitution' of bosons.

1925 – Wolfgang Pauli (25 years old) published the famous exclusion principle, later named the 'Pauli exclusion principle'.

1925 – Werner Heisenberg (23 years old) formulated matrix mechanics, thus giving birth to quantum mechanics.

1926 – Erwin Schrödinger formulated wave mechanics and published his famous Schrödinger equation.

1926 – publication of Fermi-Dirac statistics, the 'constitution' of fermions (Fermi was 25 years old, while Dirac was 24 years old).

1927 – Werner Heisenberg published his famous 'uncertainty principle' (later named Heisenberg's uncertainty principle).

1928 – Paul Dirac published his famous relativistic wave equation, later known as the Dirac Equation.

1935 – publication of the Einstein-Podolsky-Rosen (EPR) paradox, which, together with Schrödinger's work that later followed, laid the ground for the investigation of quantum entanglement.

The two theories—quantum mechanics and relativity theory—are today the most fundamental theories. The present day's whole understanding of the universe and reality rests on these two theories.

While Franz Joseph Haydn, Wolfgang Amadeus Mozart and Ludwig van Beethoven form the Viennese School, Arnold Schoenberg and his students Alban Berg and Anton von Webern form the New Viennese School (sometimes also called the Second Viennese School). Although Schoenberg is usually credited as the one who discovered the possibility of composing music beyond classical tonality and as the inventor of dodecaphony, in that era there were many composers who discovered the 'outer space' of music beyond classical tonality and invented their own new systems of structuring pitch-relations in music. It is plausible that post-tonal world and post-tonal systems were respectively discovered and invented by composers across the Europe: Alexander Nikolayevich Scriabin (1872–1915), Nikolai Andreevich Roslavets (1881–1944), Nikolai Borisovich Obukhov (1892–1954) and Arthur-Vincent Lourié (1892–1966) from Russia; Josef Matthias Hauer (1883–1959) and Fritz Heinrich Klein (1892–1977) from Austria, and some other composers (Abdyssagin 2022, pp. 12–18, 20–25).

Arnold Schoenberg (1874–1951) was a composer, teacher, musical intellectual and innovator, discoverer and creator of dodecaphony, twelve-tone technique. He proposed specific and clear rules for writing new music based on tone rows—12 non-repeating equal tones—thereby laying the foundations for all contemporary classical music. At the same time, in relation to rhythm structures and textures, he relied on the great predecessors and masters of the past, thus Schoenberg was a spiritual heir to Viennese classical and romantic traditions. He clearly demonstrated how changing, rethinking and 'modernising' the tonal foundation can radically change the perception and essence of musical art.

As Napoleon once said, 'To understand a man, look at the world when he was twenty' (Roberts 2018, p. 32). From this point of view, the era and time of Arnold Schoenberg itself, in one way or another, perhaps even subconsciously, had an incredible impact on his way of thinking and worldview.

The First World War (1914–1918) dramatically changed the world, and Schoenberg himself served in the Military service at WWI. The four largest empires collapsed (disintegrated)—German, Austro-Hungarian, Ottoman and Russian. The first socialist state in the history of mankind, the USSR, was born. The Spanish flu pandemic occurred, a horrific disaster that took the lives of millions of people, among whose victims were the artist Gustav Klimt, the revolutionary and Chairman of the All-Russian Central Executive Committee (officially the head of state of the Soviet state) Yakov Sverdlov, Major General Leonid Kapitsa (father of the future Nobel Prize winner, physicist Pyotr Leonidovich Kapitsa), and many others. Later, in 1933, the Nazis came to power in Germany (as a result of which Schoenberg had to emigrate to the USA), the Second World War (1939–1945) began and ended, and the picture of the world changed radically. At the same time, it was a period of enormous growth in science and universal discoveries, which revolutionised the understanding of the structure of the universe, thanks to which the entire civilisation entered a new stage of its development. An era that is highly contradictory, where on the one hand, there are unimaginable cataclysms, broken destinies and crippled generations of people, and on the other hand, unstoppable progress and development. And all this happened within just 50 years. A shock the like of which has never been seen before. All this, along with the whole world, was witnessed by Schoenberg, who can rightfully be called the prophet of a new era in music (Fig. 8.1). Schoenberg's work went through several important stages: from late German romanticism and expressionism to dodecaphony—and these are just the main ones. It is significant that at the end of his life he combined all the techniques. One example of the interweaving of different directions in Schoenberg's work is his Chamber Symphony No. 2, Op. 38, which was begun in 1906 and was completed in 1939.

It is no coincidence that the artist Edward Hopper said, 'Great art is the outward expression of an inner life in the artist, and this inner life will result in his personal vision of the world'.

Schoenberg's music and his concepts are still perceived ambiguously. At the time of creation his music was beyond perception for many. In a certain way, the way a person perceives Schoenberg's music can serve as a measure of that person's intellectual development.

Fig. 8.1 Leopold Godowsky, Albert Einstein, Arnold Schönberg, Carnegie Hall, 1 April 1934. Photo by Clyde Fisher (Arnold Schönberg Center, Wien [ASCI PH1473])

During my trip to Austria in 2011 I visited the Vienna Central Cemetery where I was able to pay homage to the grave of the great Arnold Schoenberg, along with the graves of Ludwig van Beethoven, Franz Schubert and other prominent composers.

An interesting element from Schoenberg's biography is that he had triskaidekaphobia. Triskaidekaphobia is a fear of the number 13. Schoenberg was born on 13th September and died on 13th July. Triskaidekaphobia is the reason why the title of Schoenberg's last opera is *Moses und Aron*; the correct name should have been *Moses und Aaron*, but that would result in a title with 13 letters, and Schoenberg wanted to avoid by all means having anything related to the number 13, so he shortened the name Aaron to be Aron, and finally the title is *Moses und Aron* which is comprised of 12 letters. Schoenberg always feared the age when he would turn 76, because $7 + 6 = 13$. Ironically and tragically, Schoenberg died at the age of 76. Moreover, according to legend, on July 13, 1951, he walked tense all day, and his last word was 'harmony'. Interestingly, Franklin Delano Roosevelt, the 32nd President of the US, also had a mild case of triskaidekaphobia (Perry, no date).

Many great pianists, including Glenn Gould and Maurizio Pollini, in addition to having mastered almost the entire established and traditional piano repertoire, had a deep respect for contemporary classical music. There are recordings of works by Karlheinz Stockhausen performed by M. Pollini, and the great G. Gould released a number of CDs where he recorded the most significant works of the New Viennese School, including Piano Sonata Op. 1 by Alban Berg, *Variations* Op. 27 by Anton Webern and piano pieces by Arnold Schoenberg.

Each member of the Second Viennese School had its own unique individuality. If Schoenberg is considered the founder of this direction, then Alban Berg (1885–1935) held an amazing balance between expressionism, late romanticism and the twelve-tone technique. He enriched the twelve-tone technique with the world of expressive symbolism and an emotional aesthetic paradigm.

About Anton Webern (1883–1945)—another of the 'holy trinity' of the Second Viennese School—was said that that he was 'plus royaliste que le roi' (more royalist than the king), in the sense that he went even further than his teacher Schoenberg, and the dawn of the technique of serialism is associated with the name of Webern. Unfortunately, he did not compose many works, but each of his music pieces is a pearl of the world musical treasury.

Schoenberg wanted to show that rethinking tonality leads to a radically different phonic results. But the contradiction between the new tonal system (and dodecaphony is exactly a new tonal system) and the old means of musical expression still existed, and this was a hindrance to the development of new music. Webern abandoned the old means and used his own, already new texture, new rhythmic structures etc. And this was a colossal breakthrough into the future.

In that era there were a number of other trends and directions that cannot be ignored. In particular, neofolklorism was a prominent direction in the first third of the twentieth century. Its most famous representatives were Béla Bartók and Zoltán Kodály.

Béla Bartók (1881–1945) was a Hungarian composer who perfectly understood all the discoveries and innovations of the New Viennese School, but implemented them in his own way. His compositional language organically combined and mutually enriched folk elements with advanced techniques of new music. Some of his well-known compositions include *Music for Strings, Percussion and Celesta*, Sz. 106, BB 114 (1937), *Mikrokosmos* in six volumes, Sz. 107, BB 105 (1940), two volumes of short piano pieces *For Children* (1945b), *Concerto for Orchestra*, Sz. 116, BB 123 (1945a) and others.

Edgard Varèse (1883–1965) was an influential composer who made substantial contributions to the practical and theoretical ontology of timbre and rhythm in music, especially in the context of an orchestral texture. Varèse and his music formed a new ideology and metaphysics of sound. Yuri Kasparov considers Varèse to be the founder of the direction of neoarchaism. There is a great deal of interesting peculiarities and details in the relations between Varèse and Webern (both were born in the same year 1883). The two composers had radically contrasting views and methods, but ultimately their creative activities occupied almost completely different domains and spaces of music: while Webern was creating perfect pitch-structures and systems with twelve-tone technique, Varèse discovered the inner universe of sound and timbre. Allegorically speaking, Webern 'controlled everything', Varèse 'controlled nothing'. However, by 'controlling everything' Webern in fact 'controlled nothing', and when Varèse 'controlled nothing', he in fact has in hold of 'everything'. Although fundamentally different, the approaches of these two composers laid the background for the developments of contemporary music in the second half of the twentieth century.

In the 1950s three of the greatest composers of their time came together— Pierre Boulez, Karlheinz Stockhausen and Luigi Nono—to form what would later be called 'the Darmstadt three'. In 2012 Yuri Kasparov proposed the idea that within 'the Darmstadt three' Luigi Nono had the same place that Alban Berg had in the 'New Viennese three', as concepts and music of Nono and Berg added new ideological and spiritual dimensions to the techniques cultivated by other members of the respective groups. Quite similarly, a certain analogy can be drawn between the approaches of Webern and Boulez.

Along with this, two titans stand apart in the history of music of the twentieth century—Igor Stravinsky (1882–1971) and Olivier Messiaen (1908–1992). Their greatness and contribution to music are so extensive that their very names have already become symbols of new time and new discoveries in music.

Stravinsky, in principle, can be considered one of those who predetermined the fate and path of development of music and the ideology of the arts in the twentieth century.

Speaking about the most iconic works of Messiaen, *Vingt Regards sur l'Enfant-Jésus* (1944), *Turangalîla-Symphonie* (1948), *Quatre Études de rythme* (1950), *Oiseaux exotiques* (1956), *Catalogue d'oiseaux* (1958) etc. are vivid examples of his genius. One of Olivier Messiaen's most expressive works is undoubtedly the *Quatuor pour la fin du temps* (Quartet for the End of Time) composed in 1941. As a soldier at World War II, Messiaen was captured by the German forces and imprisoned in a prisoner-of-war camp Stalag VIII-A, located in Görlitz, Germany (now Zgorzelec, Poland). There Messiaen composed *Quatuor pour la fin du temps*. A clarinetist, a violinist, a cellist, and Messiaen himself as a pianist, were prisoners in this camp, so this fact determined the instrumentation of the quartet (clarinet, violin, cello, and piano). The premiere of this work took place there.

Messiaen was also a prominent teacher. Among his students were Pierre Boulez, Karlheinz Stockhausen, Iannis Xenakis, György Kurtág, Gérard Grisey, Tristan Murail, George Benjamin and a number of other distinguished composers.

Karlheinz Stockhausen (1928–2007) was a unique personality in the world of music. I believe that he gave the key impetus to the development of musical art in the third millennium. A man who, having comprehended and brought to its apogee the art of serialism, moved on to a completely new art—a conglomerate of theater, painting, music, light, religion. The late works of Karlheinz Stockhausen are the gateway to a completely new musical universe. As I wrote in 2017, 'if Johann Sebastian Bach is the Isaac Newton of music, then Karlheinz Stockhausen is its Albert Einstein'.

Pierre Boulez (1925–2016) was a composer whose name is equal to that of the entire century. In one of his interviews in 2016 (when Boulez died), Ivan Fedele—the leader of the musical avant-garde of Italy—said that 'con la morte di Boulez finisce il XX secolo musicale' (with the death of Pierre Boulez, the twentieth century in music ended). One of the prominent composers of modern Russia, Yuri Kasparov, after the death of Boulez wrote an article titled 'The King died, an era ended' (2016). The name of Pierre Boulez, his very personality, is the greatest phenomenon in the music of the twentieth century.

Luigi Nono (1924–1990), along with Bruno Maderna (1920–1973), Luciano Berio (1925–2003) and others, made a huge leap in the development of Italian music. Works by Nono such as *'Hay que caminar' soñando* (1989), *Il canto sospeso* (1956), *Polifonica–Monodia–Ritmica* (1951), *A Carlo*

Scarpa, architetto, ai suoi infiniti possibili (1984), etc. are amazing discoveries in the world of music and art.

During my doctoral studies (Master di II Livello) at Conservatorio G. Verdi di Milano, Maestro Gabriele Manca analysed Nono's micro-intervallic orchestral work *A Carlo Scarpa, architetto, ai suoi infiniti possibili* (the title is a dedication to the architect Carlo Scarpa). According to Manca, everything in this piece is built on variations of two notes. It has already been noted that the spectrum of sounds is continuous. In Luigi Nono's *A Carlo Scarpa*, there is a variation on two notes: F and E♭, and the variations are huge—from microtones around E♭ all the way to E♮. That is, from E which is closest to D, to E which is closest to F. The same approach used with the note F. Manca provided arguments on whether Nono had a series and how he created dynamics in this work. In the structure of this work, in particular, there is a certain series (a row) of 12–13 different dynamic instructions.

There is one characteristic episode related to the work of another great Italian composer Luciano Berio. During my doctorate (*Corso di Perfezionamento*) in composition at Accademia Nazionale di Santa Cecilia in Rome, Italy, Ivan Fedele, who was closely acquainted with Berio, told us a story: Many years ago one of the students asked Berio a question: 'Maestro Berio, your symphony—the first chord is so beautiful, how did you make it?' Then, Berio, as if not understanding the question, replied: 'What chord? It's the timbre!'. This is proof that Berio thought music not as a set of pitches, but as a set of colours, that is, timbres. Berio's attitude towards timbre reveals that he perceived the orchestra as a single global structure, as a kind of living, constantly changing organism.

John Cage (1912–1992)—was a true phenomenon not only in America, but in the whole history of music. He was a student of Schoenberg (when Schoenberg was based in California, US), and later changed the very understanding of what the music is. One of Cage's most famous works is 4′33″— four minutes and thirty-three seconds of silence. This three-movement work was written for any instrumental combinations and can be performed by anyone, anywhere.

An example of how a pianist performs it: he goes on stage, sits down at the piano and, with the precision of a chronometer, counts for 4 min 33s, gets up and leaves. There are many interpretations of this work; some find parallels between 4′33″ by Cage and the *Black Square* by Kazimir Malevich and the works of Alphonse Allais. One of the interpretations of 4′33″ says that 4 min 33s is the time that a person can simply wait for something without switching attention. 4 min 34s—the person is already distracted, 4 min 32s—the person has not yet reached the point of tension. This, of course, is subjective, but the psychological effect is such that when the pianist, having 'performed' this piece, gets up from the piano, the whole audience sighs because the tension has gone away. Cage also has a piece *Waiting* for piano, where the idea of tension and subsequent relaxation is more accurately expressed.

Apart from that, the usually accepted interpretation is that during the performance of 4′33″ listeners should listen to the sounds that arise in the environment during the 'silence of performance'. As Cage famously said, 'There's no such thing as silence'.

Another interpretation draws an analogy with science. 4′33″ = 273 seconds, and minus 273 degrees Celsius (−273.15 °C) is the point of absolute zero! According to the third law of thermodynamics, absolute zero is the point when all motion stops. While at a temperature of absolute zero all motion ceases, quantum mechanical Zero-point energy is still retained.

4′33″ was premiered on 29th August 1952, by American pianist David Tudor. There is a legend that someone from the audience was shocked and utterly critical about 4′33″ and said, 'anyone could have composed this!'. To this Cage replied: 'Yes, but no one did it before'. It is noteworthy that this work is available for sale both as a score and as an audio performance. 4′33″ is one of the very few pieces of contemporary music that can be played literally by anyone. Sometimes people perform 4′33″ without even noticing it.

History has its own rules, and when the need arises for a certain idea, a person to realise that idea will come. The understanding of what, in principle, musical art can be, drastically changed in the early twentieth century. The very phenomenon of sound and the pitch ceased to play the dominant role. If before the music developed in two dimensions—pitches and rhythms, which were the principal and system-forming co-ordinates—then in the twentieth century the timbre-texture co-ordinate emerged as equal to the co-ordinates of pitches and rhythms.

One of the personalities who greatly contributed to the development of the concept of timbre, without whom it is impossible to imagine the world of contemporary music, is the German composer Helmut Lachenmann (born in 1935).

The fundamental essence and innermost basis of Helmut Lachenmann's music is built on the interaction of instruments in a new quality. His music is often called *musique concrete instrumentale* or even 'noise music' when sounds are performed on the instruments not in a traditional way, but mostly using completely new and previously unimaginable techniques. It is not only about the preparation of piano (which was done previously by John Cage, George Crumb and a vast number of other composers), but also about a fundamentally new approach to understanding the essence of instruments. After all, it is one thing to simply place a foam under the strings of a violin in order to hear what result will it produce—some avant-garde composers did this—but a totally different case if a technique is presented as a concept embodying and transmitting new artistic ideas. Lachenmann's music, be it string quartets, pieces for solo piano or large-scale orchestral and theatrical works, are compositions in which every technique (relatively speaking every noise or rustle) appears not by chance but in a strictly constructed, logically scrupulous system with an ideal balance of forces.

When mentioning the balance of forces, the timbral characteristics of the instrument are meant. Even in the era of classicism, W.A. Mozart's Gran Partita—Serenade No. 10 for winds in B-flat major, KV 361/370a is a kind of theater of instruments; and approaches acknowledging these unique qualities of instrumentation were largely inherited by the composers of the twentieth century. In the case of Lachenmann's musical world, not only is each technique 'clothed/dressed' in artistic integrity, but also entire cascades of techniques, certain spatial sound combinations have an enormous 'gravitational force', changing everything else around them, literally subjugating the dramaturgy of musical development. One of the most unique aspects of his compositional 'formal language' can be seen in this feature.

To appreciate a phenomenon of new music it is crucial to understand its origins—why it appeared. By analysing the evolutionary path of music one can come to the understanding that nothing happened accidentally. Even, for example, the elements of the 'Furniture music'—a term coined by the influential and eccentric French composer Erik Satie—had a lasting impact on further advancements of musical language.

A correct and complete understanding of contemporary classical music is impossible without the figure and music of Lachenmann. The modern Russian composer Vladimir Tarnopolski even called Lachenmann 'the Beethoven of our time'.

Hans Werner Henze (1926–2012) can be mentioned among the important personalities of German contemporary music. He is noted for the extremely vast range of styles and techniques implemented in his works.

Another key figure who essentially determined the development of new music is Brian Ferneyhough (born in 1943)—the founder of the idea and the whole ideology of the 'new complexity' school of composition.

Polish composer Witold Lutosławski (1913–1994) made a huge impact on the development of sonoristics, aleatoric technique and orchestral textures. His greatest influence was in the field of symphonic music: large-scale works such as *Livre pour orchestre* (1968), *Concerto for Orchestra* (1954),

Concerto for Cello and Orchestra (1970), *Concerto for Piano and Orchestra* (1988), *Symphonies* No. 1–4 (1947, 1967, 1983, 1992) are among the most significant compositions written in the twentieth century in general.

Considering the Italian composition school, in the musical era that came after Gian Francesco Malipiero (1882–1973) and Ottorino Respighi (1879–1936), the most prominent figures were Luigi Dallapiccola (1904–1975)—a dodecaphonist who continued the ideas of the great Schoenberg—and Goffredo Petrassi (1904–2003). In general, the entire history of art closely correlates with the history of the formation of Italy itself. Italian culture is a great and inseparable part of the European world, and without the great European world it is impossible to imagine the entire human civilisation. Just a simple mention of such names as Cimabue, Giotto, Piero della Francesca, Sandro Botticelli, Leonardo Da Vinci, Michelangelo Buonarotti, Raffaello Sanzio, Tiziano Vecellio, Tintoretto, Veronese, or the great literary figures Dante, Petrarch, Boccaccio, Tasso, Manzoni, Eco display the profound impact of Italian arts and culture.

Franco Donatoni (1927–2000) and Salvatore Sciarrino (born in 1947) are the iconic figures of twentieth-century Italian contemporary music. The fundamental differences between their compositional languages lie in what they consider to be the minimum musical 'building' element.

The conceptions and attitudes towards the minimum musical 'building' element largely determine the music as a whole. For example, in case of the great Italian composer Giacinto Scelsi (1905–1988), who was nicknamed 'il re di una nota sola', that is, 'the king of one note', because he composed influential works of music based just on a single tone. So often a single tone was the minimum musical element for him. According to the current view of numerous Italian composers of today, the minimal element of both Donatoni and Sciarrino was the 'musical figure'. They had an enormous influence on the development of all European music. There is now a whole generation of the so-called 'Sciarrini' (intellectual, artistic and spiritual followers of Sciarrino), while many of Donatoni's followers were nicknamed 'Donatini'.

After Donatoni and Sciarrino, a generation of other Italian composers appeared who are now shaping the musical face of Europe, in particular Ivan Fedele, Alessandro Solbiati, Sandro Gorli, Stefano Gervasoni, Fabio Vacchi, Giorgio Battistelli, Pierluigi Billone, Gabriele Manca, Giorgio Netti, Giorgio Colombo Taccani, Giuseppe Giuliano, Francesco Filidei, Oscar Bianchi, Clara Iannotta, and others. Fausto Romitelli (1963–2004) was a bright representative of Italian new music. His cycle *Professor Bad Trip: Lesson I, II, III* has been inspiring the younger generations of composers.

When considering the Austrian composition school, Beat Furrer, along with Klaus Lang, Clemens Gadenstätter, Olga Neuwirth, Matthias Kranebitter and others form the face of modern Austrian music. Among the patriarchs of modern Austrian music Friedrich Cerha and Kurt Schwertsik can be noted.

Speaking of France, another great center of European civilisation, the mainstream movement in music (after the age of Boulez) is the technique of spectralism. The spectral technique is represented by three distinguished composers—Gérard Grisey (1946–1998), Tristan Murail (born in 1947) and Hugues Dufourt (born in 1943). The pivotal works of these composers include *Vortex temporum* (1995) and *Quatre chants pour franchir le seuil* (1998) by Grisey; *Treize couleurs du soleil couchant* (1978) and *Gondwana* (1980) by Murail; and *L'Origine du monde* (2004) by Dufourt.

The spectral technique is a technique that concentrates on sound spectra (from where its name comes), and uses mathematical analysis of sound spectra (commonly with the aid of computers). It is vital to emphasise the term spectral *technique*. As Tristan Murail always says, 'there is no spectral music or spectral style, there is only spectral technique'. Another essential position of Murail is that 'music is an architecture in time'.

The question of who should be classified as spectral composers is an ongoing debate in musicology. In 2013 during the Impuls Academy in Graz (Austria), Georg Friedrich Haas even gave a special lecture titled 'I am not a spectral composer'.

Following the spectral technique, musical 'saturation' became a major movement in France, represented by Franck Bedrossian, Raphaël Cendo and Yann Robin.

Unfortunately, it is impossible to include all marvelous, brilliant and distinguished composers of avant-garde music. Thus, this historical overview serves as a means of encouraging the reader to delve deeper into the wonders of new music.

References

Abdyssagin, R.-B. 2022. *Noveishie kompozitorskie i ispolnitel'skie tehniki* (The Newest Composing and Performing Techniques). Astana: KazNUA.

Bartók, B. 1937. *Music for Strings, Percussion and Celesta*, Sz. 106, BB 114. Vienna: Universal Edition.

Bartók, B. 1940. *Mikrokosmos* in Six Volumes, Sz. 107, BB 105. London: Boosey & Hawkes.

Bartók, B. 1945a. *Concerto for Orchestra*, Sz. 116, BB 123. New York: Boosey & Hawkes.

Bartók, B. 1945b. *For Children* in Two Volumes.

Berg, A. 1910. *Klaviersonate* Op. 1. Berlin: Schlesinger.

Cage, J. 1952. *4'33"*. Edition Peters.

Cage, J. 1952. *Waiting* for Solo Piano.

Dufourt, H. 2004. *L'Origine du monde*. Paris: Éditions Henry Lemoine.

Grisey, G. 1995. *Vortex temporum*. Paris: Ricordi.

Grisey, G. 1998. *Quatre chants pour franchir le seuil*. Milan: Ricordi.

Lightman, Alan P. 2005. *The Discoveries: Great Breakthroughs in 20th-Century Science, Including the Original Papers*. Toronto: Alfred A. Knopf Canada.

Lutosławski, W. 1947. *Symphony No. 1*.

Lutosławski, W. 1954. *Concerto for Orchestra*.

Lutosławski, W. 1967. *Symphony No. 2*.

Lutosławski, W. 1968. *Livre pour orchestre*.

Lutosławski, W. 1970. *Concerto for Cello and Orchestra*.

Lutosławski, W. 1983. *Symphony No. 3*.

Lutosławski, W. 1988. *Concerto for Piano and Orchestra*.

Lutosławski, W. 1992. *Symphony No. 4*.

Mozart, W.A. 1782. *Serenade No. 10 for Winds in B-flat Major*, KV 361/370a *Gran Partita*.

Messiaen, O. 1941. *Quatuor pour la fin du temps*. Paris: Éditions Durand.

Messiaen, O. 1944. *Vingt Regards sur l'Enfant-Jésus*.

Messiaen, O. 1948. *Turangalîla-Symphonie*.

Messiaen, O. 1950. *Quatre Études de rythme*. Paris: Éditions Durand.

Messiaen, O. 1956. *Oiseaux exotiques*.

Messiaen, O. 1958. *Catalogue d'oiseaux*.

Murail, T. 1978. *Treize couleurs du soleil couchant*. Paris: Éditions Musicales Transatlantiques.

Murail, T. 1980. *Gondwana for orchestra*. Paris: Éditions Musicales Transatlantiques.

Nono, L. 1951. *Polifonica–Monodia–Ritmica*.

Nono, L. 1956. *Il canto sospeso*.

Nono, L. 1984. *A Carlo Scarpa, architetto, ai suoi infiniti possibili*.

Nono, L. 1989. *'Hay que caminar' soñando*.

Perry, W. n.d. Fears of the Fearless FDR: A President's Superstitions for Friday the 13th. Smithsonian Institution. https://npg.si.edu/blog/fears-fearless-fdr-president's-superstitions-friday-13th (Accessed 30 December 2023).

Roberts, A. 2018. *Churchill: Walking with Destiny*. Penguin Books.

Romitelli, F. 1998–2000. *Professor Bad Trip: Lesson I, II, III*.

Schoenberg, A. 1932. *Moses und Aron*, Oper.

Schoenberg, A. 1939. *Kammersymphonie No. 2*, Op. 38.

Shostakovich, D.D. 1941. *Symphony No. 7* in C Major, Op. 60, 'Leningrad'.

The University of Chicago. 1896. *The Annual Register: July, 1895–July, 1896*. Chicago: The University of Chicago Press.

Webern, A. 1937. *Variations for Piano*, Op. 27. Vienna: Universal Edition.

Wells, J.D. 2016. Prof. von Jolly's 1878 prediction of the end of Theoretical Physics as Reported by Max Planck. *Scholardox E7*. Ann Arbor: The University of Michigan Library.

Part II

Heisenberg's Uncertainty Principle and Aleatoric Technique in Music

9

Entropy in the Development of Tonality

In physics, according to the Second Law of Thermodynamics, the total entropy of a closed system either always increases or remains constant, it can never decrease. According to Claude Shannon's information theory, 'the system reaches its death when it contains as much information as it can handle' (Vedral 2018, p. 58), so no new information can be stored in it.

Here an analogy with the development of tonality can be found. In the classical or traditional sense tonality was a system with its own laws of regulating interrelations between musical intervals. From the end of the era of baroque and until the very beginning of the twentieth century tonality used to extend to become the main driving force behind the laws of composition regulating relationship between not only the intervals but also influencing all other parameters of music as well. Tonality was the dominant system that used to form the framework of a harmony. Indeed, tonality was actually harmony, or, to say better, tonality is a system for embodying harmony.

Nevertheless, according to the opinions of numerous composers and scholars, at the end of the nineteenth century the situation in the world of classical music was such that there could no longer be any single new and/or fresh melody that would not repeat something that already exists. Anything (any melodic or harmonic element) within classical tonal system would inevitably resemble something already existing, and would appear as merely new embodiments of already existing combinations of sounds. Speaking in terms of entropy, the system reached its maximum capacity of handling new information. The very impossibility of the existence of new melodic combinations, an intervallic interactions within the established system of conventional tonality has led to a crisis and collapse of the system from inside. Without

© The Author(s), under exclusive license to Springer Nature Switzerland AG 2024
R.-B. Abdyssagin, *Quantum Mechanics and Avant-Garde Music*,
https://doi.org/10.1007/978-3-031-63161-0_9

arguing about the validity of this statement, it is worth observing that the very existence of such an opinion was a highly important element in the emergence of new music. As composer Yuri Kasparov recalls, a composer, musicologist, conductor, Professor Evgeny Makarov during his lectures in the Moscow State Tchaikovsky Conservatory in 1983 always repeated that Nikolai Andreyevich Rimsky-Korsakov said that at the end of the nineteenth century and the very beginning of the twentieth century, within the established tonal system 'not a single new intonation and not a single new harmonic turn can appear'.

Since the age of Johann Sebastian Bach (1685–1750) polyphony as means of organising the structure of musical work has been eclipsed by the emerging power of a tonal system, so 'in Bach, tonality answers the question how is polyphony possible as harmonic polyphony' and 'Bach is ... a harmonist' (Adorno 2016, p. 62). This is an exceptionally fundamental point for understanding the key difference between, for example, the age of Johann Sebastian Bach, Georg Friedrich Händel (1685–1759), Domenico Scarlatti (1685–1757) and methods of any preceding polyphonist and counterpointist composers since Leoninus and Perotinus Magnus.

In Bach's music the vertical dimension (harmony) was a decisive force in formulating the horizontal (polyphony) relations between voices. Therefore, the polyphony as itself retained its richness and a certain degree of autonomy within the influential field of tonality. Then, with the help of Christoph Willibald Gluck (1714–1785), Franz Joseph Haydn (1732–1809) and others, tonality has been rapidly expanding its field of influence on the structure of musical composition.

The era of classicism was one of the key periods in the history of music. Classical music was born after the law of universal gravitation was discovered and the classical mechanics of Isaac Newton appeared (1687, year of the first publication of Newton's *Philosophiæ Naturalis Principia Mathematica*) (Figs. 9.1, 9.2 and 9.3).

It was in the era of classicism when tonality received the role of an absolute foundation, a distant analogue of the physical phenomenon of 'gravitational force' in music. At the same time the form and even the principle of sonata appeared as a reflection of the law of dialectics in the works of the Viennese classics—Franz Joseph Haydn, Wolfgang Amadeus Mozart (1756–1791) and Ludwig van Beethoven (1770–1827). Especially in Mozart's and Beethoven's œuvres tonality and the tonal system reached its zenith, an absolute apogee. Tonality became a certain kind of gravitational force attracting all other structural elements of composition (Abdyssagin 2023).

Fig. 9.1 I. Newton (1642–1727) (Portrait by Godfrey Kneller, Isaac Newton Institute)

Fig. 9.2 J.S. Bach (1685–1750) (Portrait by Elias Gottlob Haussmann, Bach-Archiv Leipzig)

Fig. 9.3 I. Kant (1724–1804) (Portrait by Johann Gottlieb Becker, Schiller-Nationalmuseum)

The philosophical movement of German idealism emerged during this period and its brightest representatives were Immanuel Kant (1724–1804), Johann Gottlieb Fichte (1762–1814), Friedrich Wilhelm Joseph Schelling (1775–1854), Georg Wilhelm Friedrich Hegel (1770–1831) and others. This was followed by the Jena Romanticism: Johann Christian Friedrich Hölderlin (1770–1843), Novalis (1772–1801), the brothers August Wilhelm Schlegel (1767–1845) and Friedrich Schlegel (1772–1829) among others.

This statement about German idealism does not imply any direct connection with music; nevertheless, it is interesting to observe that one of the most important periods in philosophy chronologically matches with one of the most important periods in music—classicism. Just like classicism established some of the foremost foundations of classical music and opened the gates to the further development and emergence of romanticism, German idealism and the work of Kant were of exceptional significance for the evolution of philosophical thought and influenced the progress of philosophy.

The sonata, like a fugue, is more than just a form, it is a principle. The sonata is a whole world, a kind of reflection or even expression in music of the dialectical law of unity and conflict of opposites. Formative and, in some aspects, ideological principles of the sonata can be characterised as a search and desire to find, see and understand disunity in unity, and unity in disunity.

Important to note interesting aspects of the terminology of Charles Rosen who even referred to sonata as 'sonata forms' in plural (1988, 1997).

Then, already in the 'structural functions' (Schoenberg 1990) that harmony plays in the works of Franz Peter Schubert (1797–1828), Robert Schumann (1810–1856) and Johannes Brahms (1833–1897), it can be

Fig. 9.4 Excerpt from the score of Richard Wagner's opera *Tristan und Isolde*, WWV 90. Tristan chord highlighted in red (Piano arrangement by Richard Kleinmichel. Milan: G. Ricordi & C. Public domain. *Source* IMSLP, Petrucci Music Library)

Fig. 9.5 Excerpt from the score of Richard Wagner's opera *Tristan und Isolde*, WWV 90. Tristan chord is present in the second bar (First edition. Leipzig: Breitkopf und Härtel. Public domain. *Source* IMSLP, Petrucci Music Library)

Fig. 9.6 R.Wagner (Photo by Franz Hanfstaengl. *Source* Wikimedia Commons)

observed how inner fluctuations of musical material start to touch the boundaries and very borders of the established tonal system. Richard Wagner's (1813–1883) works—especially the opera *Tristan und Isolde* WWV 90 (1860)—displayed the first traces of dismantling, splitting and recreating the tonal order, as shown in the so-called 'Tristan chord' and free use of chromaticisms (Sadie and Latham 1985, pp. 353–355). The Tristan chord consists of F, H, D♯, and G♯ and is part of Tristan's leitmotif (Figs. 9.4 and 9.5). Although in various incarnations (and tonal variations) had already appeared in the works of other composers before Wagner, it was Wagner's contextual use of it that predicted the upcoming tonal revolution (Fig. 9.6).

Even if very delicate, these changes became the wind that caused the storm brought by Arnold Schoenberg (1874–1951) and his disciples. It is an astonishing and thrilling fact that great physicist Werner Heisenberg was deeply aware of that changes and 'observed that in the music of the late nineteenth century, with Wagner and the French impressionists, Debussy and Ravel, the structure of classical harmonies had collapsed' (Blum 2002).

Paul Hindemith wrote that 'Tonality is a natural force, like gravity' (1942, p. 152). He was eminent in his perception of tonal elements as attracting forces, and compared not only the tonality itself, but even some of its sub-structures to gravity (1942, pp. 22, 75). Additionally, Hindemith introduced the notions of harmonic energy and harmonic centre of gravity in his interpretation of tonal systems (1942, pp. 109, 115) and stated that 'Tonal coherence, or tonal "binding", is nothing more than gravity in its most refined form' (1941, p. 91). Further investigations of Hindemith's music theory as well as his implementation of the meaning of gravity can be found in the works of Simon Desbruslais (2013, 2018), Margarita Katunyan (1982, p. 65) and many other scholars. On the whole, specific mentions of analogies with gravity appeared in a work of German musicologist Carl Dahlhaus (1968, pp. 38, 221). The idea of perceiving tonality as an analogue of gravity is important as a framework and basis for elucidating other metaphoric correlations. So, Hindemith's concepts were touched because tonality as a metaphoric gravitational force can facilitate the explanation of other possible connections with physics, but it also helps to understand the role that tonality played in music and why the emergence of twelve-tone system was an event whose significance is difficult to overestimate.

This was an introductory review that served to demonstrate the rise and fall of the tonal system which collapsed under its own weight and mass. It reached its maximum entropy and had to be burned in order to be resurrected in a form of twelve-tone technique, which subsequently evolved and opened the gates to the new world with a myriad of branches, directions and techniques of new music. As Theodor W. Adorno wrote: 'Twelve-tone technique is truly the fate of music. It enchains music by liberating it' (2016, p. 45).

There is no claim that tonality (in a broad meaning) as a gravitational force became extinct. This passage only displays that traditional and conventional forms of tonal systems became mostly obsolete as a fundamental basis of avant-garde music. Nevertheless, the tonality in a much wider and liberal sense as gravity still exists, but it should be treated as an incomparably broader and generalised concept. Moreover, distinguishing what is the fundamental gravity-like force in music nowadays vastly depends on specific compositional techniques and sometimes even varies depending on particular music work.

The tonal background of the dodecaphonic system as well as internal profound links and connections between 'old' (classical tonality) and 'new'

Fig. 9.7 First bars of Präludium from *Suite* Op. 25 by A.Schoenberg. The score was published by Universal Edition in Vienna in 1925 (© By kind permission of Universal Edition A.G., Wien)

(Schoenberg's dodecaphony) systems were brilliantly and vividly demonstrated by Philip Herschkowitz in his article *Dodekafoniya i tonalnost* (Dodecaphony and Tonality) where he conducted structural analysis of the very beginning of A. Schoenberg's *Suite for Piano* Op. 25 (Fig. 9.7). At the end, Herschkowitz arrives at an exceptionally important conclusion that 'the dodecaphonic system is in the same position of continuity with respect to the tonal system in which the tonal system was in relation to the seven-mode system of the Middle Ages' (1991).

American musicologist Jack Boss conducted outstandingly close in-depth analysis of Schoenberg's *Suite for Piano* Op. 25 in his book *Schoenberg's Twelve-Tone Music: Symmetry and the Musical Idea* (2014, pp. 35–121). Those interested in a more detailed analytical account of Schoenberg's twelve-tone technique and its implementation in Suite Op. 25 can learn a lot from that source.

Interesting to note that A. Schoenberg in the early 1950s was actively studying and researching the Volume I of J.S. Bach's *The Well-Tempered Clavier*. Using a green coloured pencil, Schoenberg attributed numbers from 1 to 12 to different noteheads in the first three bars of the score of Bach's Fugue No. 24 in B minor (Fuge h-Moll) BWV 869 (Fig. 9.8). Moreover, in this score Schoenberg made a remark: 'Is this the first composition with 12 tones?' (Arnold Schönberg Center 2023).

This example clearly shows that the very possibility of existence of music beyond the boundaries of traditional/classical tonal system was always present in music. Schoenberg, as well as numerous other composers, was one of the first to discover the possibility of composing twelve-tone music, or, in

Fig. 9.8 Das Wohltemperierte Klavier von Joh. Seb. Bach. Revidiert und mit Finger-satz versehen von Carl Czerny. Leipzig [no date] (Arnold Schönberg Center, Wien [SCO B21])

a broader sense, music beyond the conventional concept of tonality. In this sense, the possibility of writing twelve-tone music was discovered in the beginning of the twentieth century by several composers independently, as this possibility always naturally existed in music. However, the specific dode-caphonic (twelve-tone) technique was invented by Schoenberg, as the rules and methods of this particular system did not exist before Schoenberg. In fact, after discovering twelve-tone music, each of the composers/discoverers invented their own systems of working with twelve-tone music.

According to the Western 12 equal temperament system there are 12 equal tones in an octave. The key rule of the dodecaphony (twelve-tone technique) is that, within a tone row, a tone cannot be repeated until other eleven tones pass.

Music based on this or a similar system is often called 'atonal'. Never-theless, this understanding of 'atonality' is incorrect if, in a broader sense, we perceive 'tonality' as a gravity and gravitational attraction (the German word *Schwerkraft* is a suitable description) in music. 'Atonal' music is a music that has no tonality. This can be valid in a sense when music is not subject to the rules and system of classical tonality. However, if we perceive tonality as a gravity in music, then there can be almost no music without tonality (with the exception that 'atonal' music can exist in a form of amor-phous and spontaneous improvisations) because there can be almost no music without a 'centre' of the system (whatever system is used). Almost any musical direction, be it sonoristics, spectral technique, and even more so the seri-alism and the twelve-tone technique, are 'tonal' precisely in the sense that there is a foundation, a centre/focus of gravitational attraction. The only question is how this concept of tonality manifests itself because what is 'gravity' or can serve as a focus of gravitation attraction is completely different according to different musical styles and techniques. For example, in sonoris-tics (sometimes also called 'sonorism') the main focus/centre of gravitational attraction is a massive timbral-textural object that (figuratively) 'curves' all other parameters in music.

It would be useful to introduce the concept of gravitational attraction in music. And such 'gravity' exists in music of all kinds, it is not limited to the Western classical music but can be found across the world in different national traditions of music. The Western concept of 'tonality' is only one domain or a specific case of the more global and overarching paradigm of 'gravity' in music.

The emancipation and 'un-chainment' of dissonance was introduced in the music of New Viennese School. The phenomena of consonance and dissonance—as relations between tones—have existed in music since ancient times. The distinction between consonance and dissonance is very subjec-tive and depends on how a a person perceives these intervals. Consonance, relatively speaking, is an interval that is pleasant to the ear, while dissonance is not so. Which intervals (relations between tones) are consonant and which are dissonant is a highly conditional and subjective categorisation, and historically such division has depended on the specific era.

If in the Middle Ages the interval of the fourth was a dissonance that required resolution, then in the works of classical composers it is clear that the fourth is a consonance. Not to mention that the fourth is one of the main intervals in traditional Kazakh music, for example.

There is a legend that the interval of the tritone was almost banned by the church in the Middle Ages and early Renaissance. The diminished seventh chord was considered simply a 'devil's chord' because it crossed two tritones together. In contrast, in the music of Romanticism, e.g. F. Liszt's *Après une lecture du Dante: Fantasia quasi Sonata* begins with tritones, A is played in octaves followed by E♭ and so on. Thus, the tritone has become a system-forming interval in this *Dante Sonata* by Liszt.

Even the literary world felt the influence of the Second Viennese School. In particular, composer Arnold Schoenberg and his formal language in music was the prototype behind the main character Adrian Leverkühn and his musical innovations in the acclaimed novel *Doktor Faustus* (1947) by the great German writer and intellectual, Nobel Laureate Thomas Mann. *Doktor Faustus* by Mann became a hallmark literary work that can be used as a clue to understanding the richness of the inner world of a creative individual.

The influential German philosopher, social theorist and musicologist Theodor W. Adorno was an unofficial musical consultant to Thomas Mann. In his biography of Schoenberg, musicologist Mark Berry (2019, pp. 197–199) elucidates the troubles and misunderstandings that arose between Mann and Schoenberg, and primarily between Schoenberg and Adorno. Schoenberg particularly disliked Adorno for many reasons, and preferred to name him Wiesengrund rather than Adorno. Referring to Adorno's role as musical adviser for Mann's *Doktor Faustus*, Schoenberg even said that he (Schoenberg) 'would have been only too happy to have invented a special, fictional compositional system for Mann to have used instead of Wiesengrund's half-digested "twelve-tone goulash".' (Berry 2019, p. 199).

Das Glasperlenspiel (*The Glass Bead Game*, 1943) by Nobel Laureate Hermann Hesse—one of the landmark novels of the twentieth century—largely contains elements and attitudes towards the art of music that was prominent in the intellectual world of Schoenberg. Hesse's novel is one of the most iconic works in the literature. Alongside Mann's *Doktor Faustus*, this is the greatest work written about the personality of a composer and creator. Hesse's approach in building the structure of this novel and the structure of the narrative finds many twists and correlations with how Webern built the structure of his *Variations Op. 27*. I would say that *Das Glasperlenspiel* by Hesse is a symphony written in words (Abdyssagin 2022, p. 11). In addition, it is believed that Austrian composer Josef Matthias Hauer was the influence behind certain aspects of Hesse's novel.

References

Abdyssagin, R.-B. 2022. *Noveishie kompozitorskie i ispolnitel'skie tehniki* (The Newest Composing and Performing Techniques). Astana: KazNUA.

Abdyssagin, R.-B. 2023. 'About Viennese Classics' (2022), a booklet text for *Rakhat-Bi Abdyssagin plays Viennese Classics: Haydn, Mozart, Beethoven*. [CD]. Berlin: Kreuzberg Records, kr 10175.

Adorno, T. 2016. *Philosophy of Modern Music*, trans. A.G.Mitchell and W.V.Blomster. London and New York: Bloomsbury Academic.

Arnold Schönberg Center. 2023, March 16. *Composition with Twelve Tones. Schönberg's Reorganization of Music From the exhibition | Object 1*. Facebook Publication. https://www.facebook.com/100063650946680/posts/pfbid02ujyyi VFZMxLbLAHUdHT5fQ9TY3KTTTuy7Qdw7bzx9Pi7wvqNsRu4xMEy9ss25r qkl/?d=w&mibextid=uc01c0. Accessed 16 March 2023.

Berry, M. 2019. *Arnold Schoenberg. Critical Lives*. London: Reaktion Books.

Blum, B. 2002. *Heisenberg and Music*. Heisenberg Family Archive. https://web.archive.org/web/20180831103506/http://heisenbergfamily.org/hbgmusik.htm. Accessed 19 March 2023.

Boss, J. 2014. *Schoenberg's Twelve-Tone Music: Symmetry and the Musical Idea*. Cambridge: Cambridge University Press.

Dahlhaus, C. 1968. *Untersuchungen über die Entstehung der harmonischen Tonalität*. Kassel: Bärenreiter-Verlag.

Desbruslais, S. 2013. *The Identity, Application and Legacy of Paul Hindemith's Theory of Music*. DPhil Thesis. University of Oxford.

Desbruslais, S. 2018. *The music and music theory of Paul Hindemith*. Suffolk: Boydell & Brewer.

Herschkowitz, P. 1991. Dodekafoniya i tonalnost (Dodecaphony and Tonality). *O Muzyke* (On Music). Volume I. Moscow, pp. 214–247.

Hesse, H. 1943. *Das Glasperlenspiel*. Zürich: Fretz & Wasmuth Verlag.

Hindemith, P. 1941. *The Craft of Musical Composition. Book II. Exercises in Two-part Writing*, trans. O.Ortmann. New York: Associated Music Publishers, Inc., London: Schott & Co., Ltd., Mainz: B. Schott's Söhne.

Hindemith, P. 1942. *The Craft of Musical Composition. Book I (fourth edition). Theoretical Part*, trans. A.Mendel. New York: Associated Music Publishers, Inc., London: Schott & Co., Ltd.

Katunyan, M. 1982. Ponyatiye tonal'nosti v sovremennom teoreticheskom muzykoznanii (The Concept of Tonality in Modern Theoretical Musicology). *Muzykal'noe iskusstvo.*

Liszt, F. 1849. *Après une lecture du Dante: Fantasia quasi Sonata.*

Mann, T. 1947. *Doktor Faustus. Das Leben des deutschen Tonsetzers Adrian Leverkühn erzählt von einem Freunde.* Frankfurt am Main: S. Fischer Verlag.

Newton, I. 1687. *Philosophiæ Naturalis Principia Mathematica.* London: Royal Society.

Rosen, C. 1988. *Sonata Forms.* Revised. New York and London: W.W.Norton & Company.

Rosen, C. 1997. *The Classical Style: Haydn, Mozart, Beethoven.* Expanded. New York and London: W.W.Norton & Company.

Sadie, S., and A. Latham, eds. 1985. *The Cambridge Music Guide.* Cambridge: Cambridge University Press.

Schoenberg, A. 1999. *Structural Functions of Harmony. Edited by Leonard Stein.* London: Faber and Faber.

Vedral, V. 2018. *Decoding Reality: The Universe as Quantum Information.* Oxford: Oxford University Press.

Wagner, R. 1860. *Tristan und Isolde, WWV 90.* Leipzig: Breitkopf und Härtel.

Webern, A. 1937. *Variations for Piano*, Op. 27. Vienna: Universal Edition.

10

Performance as Experiment

Any scientific theory in the natural sciences has to be proven by experiment, and only after empirical approval by the means of experiment it can be considered as a law of nature. As Nobel Prize laureate, physicist Richard Feynman stated: 'The principle of science, the definition, almost, is the following: *The test of all knowledge is experiment.* Experiment is the *sole judge* of scientific "truth".' (2011, p. 2). This reveals the 'empiristic attitude of modern science'. As W. Heisenberg wrote: 'since the time of Galileo and Newton, modern science has been based upon a detailed study of nature and upon the postulate that only such statements should be made, as have been verified or at least can be verified by experiment' (2000, p. 38).

This all corresponds to the fact that music and every theory in composition also have to be proven by an experiment. The experiment is truly natural to music—it is the inseparable part of musical art. The name of this experiment is *performance*. A score or a compositional theory has to be performed, thus tested by experiment.

This field of research already has specific explanations in the work of Peter Pesic (2014). Robert P. Crease contributes to the direction called 'performance theory' and treats performance not only as a *praxis*, but also as a form of *poiesis*. He links performance in performing arts with scientific experiments in the natural sciences (1993).

In the twentieth century the number of 'theoretical' music works drastically increased, including studies and canons of Conlon Nancarrow written for mechanical piano—auto-playing musical instrument (1948–1992), or electronic music where the sounding result is already a 'performance', and compelling combinations of acoustic music with elements of electronics.

R.-B. Abdyssagin, *Quantum Mechanics and Avant-Garde Music*, https://doi.org/10.1007/978-3-031-63161-0_10

Additionally, some canons of Nancarrow are based on mathematically calculated formulas of the relation of music/thematic elements to each other.

With respect to the performance of music, there is a drastic difference between traditional scores and aleatoric contemporary scores. In short, in classical music performance alters particular details of the music, but usually to a negligible extent, and the structure of the music itself remains the same and is always recognisable. In the same way an experiment in classical physics in our everyday reality (with large objects) may change the state of an object under experiment only to a certain, and arbitrarily small, level.

On the other hand, in works of contemporary music using the aleatoric approach (such as selected compositions of Karlheinz Stockhausen, John Cage, Pierre Boulez, Morton Feldman, Bruno Maderna and others) performance changes the whole architectonics, form, shape and 'reality' of music, quite similarly to how in quantum mechanics a measurement cardinally changes the state of the object being observed; shortly, the experiment affects and changes the reality! This interesting question is reviewed in greater detail in Chapter 13.

Music truly exists in different forms, but what is a music? The score itself is not a music. As S. Covey wrote: 'we all know that "the map is not the territory." A map is simply an explanation of certain aspects of the territory' (2013). In this context the score is like a map, a description of music, but it is not the music itself. Music becomes music only when it is performed, thus, when it has been approved empirically by conducting an experiment—delivering a performance! A musical score is like a mathematical language in which physical theories are written, and only experiment can test the rightness of these theories. In the same way only performance can reveal the 'existence' of such music that had previously been 'theoretically'/imaginatively inscribed on paper.

On the other hand, there is plenty of great and highly influential music written more for 'eyes' than for 'ears'. *Die Kunst der Fuge* (The Art of Fugue) BWV 1080 by J.S. Bach seems to be an obvious example (Fig. 10.1). This magnificent, mysterious and cosmic opus—one of the most elevated and pervasive pinnacles of musical art—does not even have an indication of instrumentation. It exists purely as the 'theoretical' music score where each polyphonic voice is indicted separately. Nevertheless, another unique aspect of this work is that it can be wonderfully performed by almost any instrumentation. Scholars are still arguing on which instruments and performers this work originally was written or intended for. This cycle is not finished because of the death of the composer, and the score interrupts at a specific

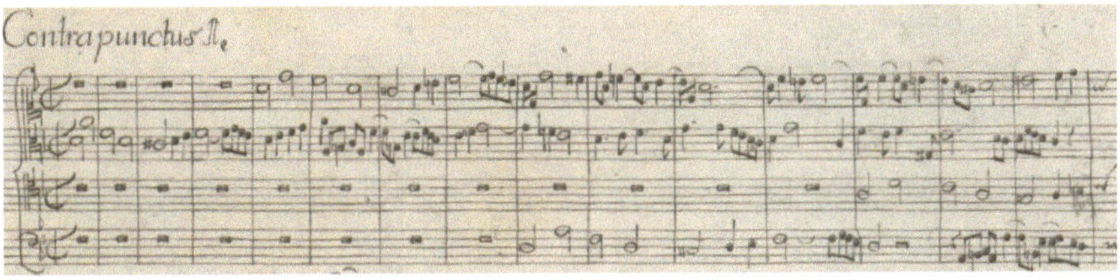

Fig. 10.1 Excerpt from J.S. Bach *Die Kunst der Fuge*, first publication by Carl Philipp Emanuel Bach (1751) (Public domain. Staatsbibliothek zu Berlin. *Source* IMSLP, Petrucci Music Library)

point of intersection of counterpoint where Bach's own surname is musically embodied (some claim it to be for the first and the last time in his music). According to the letter version of the notes, B A C H correspond to Si♭, La, Do, Si♮, and Bach himself was aware of that. So some researchers treat this is the composer's final signature written on his last and one of the most important works. The final page of the unfinished manuscript contains a handwritten note (Fig. 10.2) of Carl Philipp Emanuel Bach (son of J.S. Bach): 'Über dieser Fuge, wo der Name B A C H im Contrasubject angebracht worden, ist der Verfasser gestorben' (The composer died while working on this fugue, which introduces the name *BACH* in the countersubject). It is true that *The Art of Fugue* became a culmination of Bach's experiments with the mono-thematic variational counterpoint technique.

Since J.S. Bach the tonal combinations and variations of B A C H have been present in countless musical works, especially from romanticism to avant-garde. In particular, H C A B (the retrograde of B A C H) is present in Schoenberg's Suite Op. 25! Examples of the appearance of B A C H in dodecaphony and serialism are present in Arnold Whittall's book *The Cambridge Introduction to Serialism* (2008).

As a rule of polyphony, and precisely Bach's interpretation of imitational polyphony, voices imitate the main initial theme and then develop it, each in its own way. The genre (or compositional principle and framework) of a fugue became living embodiment not only of that principle, but also of the whole era, as polyphony is musical expression of medieval and renaissance outlooks, which has deep roots, and this question of the relation between the genre of music and society/epoch which gave it a birth has already been widely researched.

Given the contrast of the quantum world to the reality we experience in everyday life (macro-world), a number of interpretations of quantum mechanics emerged, including the 'Copenhagen Interpretation' (and Bohr's

Fig. 10.2 Excerpt from *Die Kunst der Fuge*, an unfinished score of *Fuga a 3 Soggetti* which contains C.P.E. Bach's note about introduction of B-A-C-H and death of composer (Public domain. Staatsbibliothek zu Berlin. *Source* IMSLP, Petrucci Music Library)

idea of complementarity as part of it), 'Many-worlds Interpretation', 'Everything is a Quantum Wave Interpretation' (Vedral 2023), 'de Broglie–Bohm interpretation' (deterministic, causal and nonlocal, hidden variables theory), 'Quantum Information Theories', 'Relational Quantum Mechanics' (Rovelli 1996), 'Consistent histories' (Griffiths 1984, 2003), 'Ensemble Interpretation', 'Quantum Logic' (Birkhoff and Von Neumann 1936), 'Time-Symmetric Theories' (including retrocausality) (Schottky 1921a, 1921b; Aharonov et al. 1964; Aharonov and Vaidman, 1998), 'Quantum Bayesianism (QBism)' (Caves et al., 2002)', Brukner–Zeilinger interpretation (2001), 'The Cellular Automaton Interpretation' ('t Hooft 2016), and so forth. The most prominent interpretations of quantum mechanics are the Copenhagen Interpretation, de Broglie–Bohm interpretation, Everettian many-worlds, and dynamical collapse theories.

According to the many-worlds interpretation and 'branching universe': 'whenever a quantum measurement occurs the universe branches into as many components as there are possible results of the measurement' (Rae 2018, p. 77).

Now we want to attract attention to a somewhat distant, but alluring alignment of polyphony with the 'many-worlds interpretation' of quantum mechanics. In imitational polyphony, different voices (or musical worlds) start from a single common theme, however, during the development of a form (in both linear and harmonic senses), each voice produces different versions and possibilities of how the initial theme developed further, while each voice can be considered as a separate reality.

Hugh Everett III's and David Deutsch's many-worlds interpretation and 'multiverse' may sound as 'science fiction' or fantasy, but these interpretations are among the supported interpretations of quantum mechanics. As David Deutsch declared, the Everettian many-worlds interpretation shows its full potential in the field of quantum computation.

Again, analogy with polyphony is somewhat imaginary, and has only few and fragile points of metaphoric connection. Polyphony emerged many centuries before the scientists discovered the micro-world and quantum mechanics, and there can be many more 'firmer' analogies between polyphony and other aspects of life. Additionally, there might be slightly more apparent analogy with heterophony (yet still fragile one), which can be defined as a single melody sounding simultaneously in different voices with variations.

Inspired by the idea of parallel universes, I wrote *13 Notes from the Parallel Universe* (2020) for string quartet (Fig. 10.3). It is a cycle (total duration ≈ 30 min) of 13 quasi-independent pieces for a string quartet (2 violins, viola and cello). This opus was inspired by a poetic interpretation of the idea of many-worlds and parallel universes. This is reflected in its structure: at both micro (local) and macro (global) levels. Each of the 13 pieces is independent in itself, but they all have common elements on the fundamental level: each separate piece can be considered as a musical universe, and they all show how similar 'events' evolved differently in different 'worlds'.

Nobel Prize-winning physicist Anthony James Leggett criticizes overrating the interpretations and paying too much attention and energy on formulating the interpretations, saying that '...it does not pay off to spend too much of your time to understand or interpret quantum mechanics as it is currently formulated...', and concluding that science and research itself is fundamental, and in the future we may have a much broader understanding than we have now (The UIUC Talkshow 2023, 24 min. 06s.).

There is also a view 'shut up and calculate', which expresses approach of some physicists towards interpretations of quantum mechanics. Nowadays a new phrase has became popular: 'shut up and contemplate'.

Having said that, some interpretations should not be confused or mixed up with explanations. The birth of an idea and its development into a conjecture is one of the prerequisites of making science. As D. Deutsch declared, explanations are fundamental for science. In fact, science can be viewed as a process of finding better and better explanations (1997, 2011).

Fig. 10.3 Excerpt from the score of Rakhat-Bi Abdyssagin, *13 Notes from the Parallel Universe* (2020) (© Rakhat-Bi Abdyssagin (2020). All Rights Reserved)

As a demonstration of the importance of measurement/performance, here I provide analysis (2022, pp. 96–100) of my music composition *The Space of Resonance* for 7 performers (2018) (Fig. 10.4).

In this work seven musical instruments are seven different worlds, each of which flickers in its own way and, thus, in a special way influences all the others. When the object's proper time flows extremely fast, there is a feeling that everything around it freezes and stops. Thanks to the resonance, everything revives and builds up some specific space, generated by the system of interactions of seven different worlds. The world premiere was held on 14th February 2019 in Graz, (Austria), György-Ligeti-Saal, (Das Haus für Musik und Musiktheater), Universität für Musik und darstellende Kunst, during the 11th international Impuls Ensemble & Composers Academy for Contemporary Music, performed by ensemble zone expérimentale Basel, conducted by Mike Svoboda.

In a certain sense, this work became a manifestation of the idea of complete independence of each instrument. If the part of each instrument from this score will be isolated in the form of a separate piece—it will evidently have its own dramaturgy, logic of development and formation of form, and will be complete and holistic. Each instrument is like a separate world and space, and interacting together, they create a special 'space of resonance'.

Fig. 10.4 Beginning of the score of Rakhat-Bi Abdyssagin *The Space of Resonance* (2018) (© Rakhat-Bi Abdyssagin (2018). All Rights Reserved)

The phenomenon of resonance is one of the key concepts in musical art. It has many analogies in both the physical world and domain of the spirit/ idea/ *Geist* and imagination. Touching upon the structure of the formation of a given work, it is virtually impossible to divide it into different stages/phases of the development of the form. On the one hand, with the independence and autonomy of each instrument, they all together form a single 'organism' that creates music.

The indicated tempo (by metronome, ♩ = 19–27) is extremely slow, and small durations are used deliberately, not by accident. The point is that this makes is possible to notate and control literally every fraction of a second. Accordingly, this requires an enormous concentration of performers and the skill of a conductor. Moreover, this method of notation is used to make each instrumentalist feel as comfortable as possible, and everyone can be in complete freedom. From a performing point of view, there is a feeling that everyone can decide not to listen to what the other is doing, and play their own parts, while overall there is a feeling of the integrity of the work.

The logic is that each of the seven instruments influences the other instruments by what it plays without noticing it. It is like communicating vessels, like that space in which every slightest action is repeatedly reflected on other scales. Each note, or even each pause of a single instrument, is simultaneously both a response, a 'reaction' to what other instruments had, and an 'impulse' for future effects.

One of the most effective techniques was also used when the 'horizontal' becomes 'vertical'. For example, in a 'large area' with a developed melodic (horizontal) line, all its elements (pitch/frequencies) become the sounds of a 'chord' (mega-cluster), thereby being transformed into 'vertical' (an analogue of harmony). The reverse process is also used, when the notes of a chord, according to different principles of the organisation of musical material, become elements of a pitch line, that is, harmony turns into melody. All rhythmic structures and rhythmic formulas (whether they are complex tuplets or even tuplets under tuplets) are constant.

There are many elements of both imitational and contrasting development. There are also elements of canon and various other polyphonic techniques. Some selected features of pointillism and serialism are retained. In particular, tone row and series technique is sometimes used to calculate sound frequencies at the macroform level. The 'impulse-response' principle is also featured, when one event gives rise to many others, which is, actually, one of the meanings of resonance.

Many extended methods for playing instruments have been applied. For example, half noise / half tone on the alto saxophone, elements of bisbigliando and multiphonics on the oboe, interplay with changing the position of the bow on the violin (glissando from the position 'on the bridge' to ordinario), etc.

The score is transposed. It is worth noting the frequent change in the rhythmic size of the bars. On the one hand, this eliminates and prohibits rhythmic monotony, on the other hand, a special 'polyphony of rhythms' and the game (interaction) of polyphonies of 'intersecting spaces' are created. In addition, this approach can be considered as a reflection of the 'beats' and (un)uniform waves of 'resonance' propagation in different environments of space.

At the same time, ancient polyphonic techniques such as inversion, retrograde, inversion of retrograde (and vice versa) and many others have been implemented. These techniques were applied at the level of macroform and global structuring of form and space. The above-mentioned techniques are used not only for individual passages or series, but for an entire section of a musical work. While having plenty of pauses, the music almost never stops.

Bar 2/ marks a transition to the key section of this work. Here, using a concrete example, we see the acoustic implementation of the idea of the space of resonance.

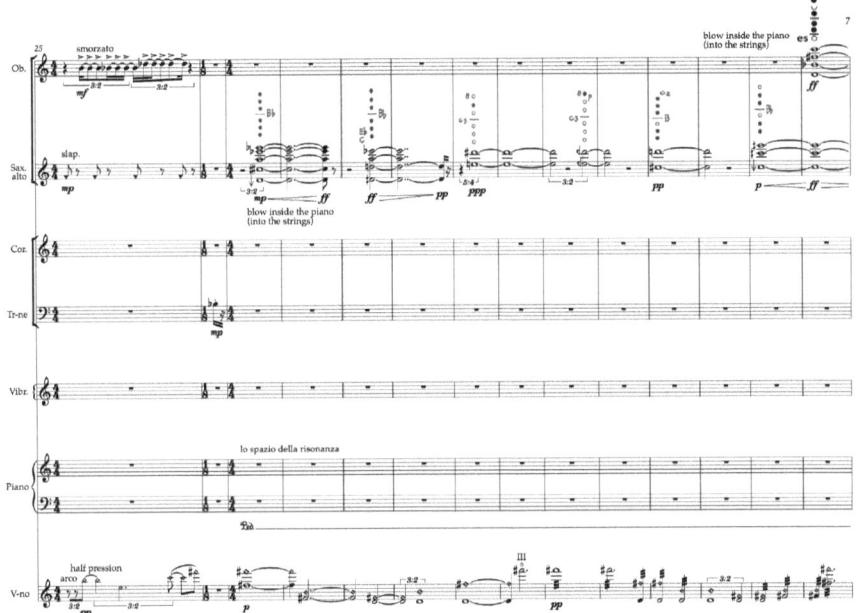

Fig. 10.5 Excerpt from the score of Rakhat-Bi Abdyssagin *The Space of Resonance* (2018) (© Rakhat-Bi Abdyssagin (2018). All Rights Reserved)

Let's pay attention to the violin part (Fig. 10.5). It plays different harmonics, but the acoustic result of these different harmonics produces the same sound (F♯$_6$). This give rise to an analogy with the bisbigliando technique, but on a string instrument.

In this section, the saxophone plays multiphonics in different dynamics (Fig. 10.5). At the same time, the 'space' of all other instruments is specially 'freed up'. In this phase, the violin acts as a kind of 'constant', playing the role akin to that of an 'orchestral pedal'.

The exceptional acoustic properties of a grand piano play a substantial role here. It is no coincidence that 'lo spazio della risonanza' (the space of resonance) is written over the piano part. The pianist does not play anything himself but presses and holds the pedal. This 'opens' the piano strings and makes them even more sensitive to the smallest changes in the acoustic environment. The saxophonist plays the multiphonics 'inside' the piano, as close as possible to the side of the piano strings. Thus, the sound waves produced by the saxophone enter into acoustic resonance with the corresponding strings of the piano. The pianist only holds the pedal. The strings of the piano begin to sound themselves. An obvious resonance phenomenon occurs as the strings resonate with the sounds of the multiphonics from the saxophone. Multiphonics are different, and each multiphonic has its own, very special and unique 'resonance' field. Towards the end, an oboe joins with its multiphonic.

The end of this period and a very significant musical phase is the acoustic technique of a 'sound shot' when the (wah-wah) mute on the trombone is sharply removed (Fig. 10.6).

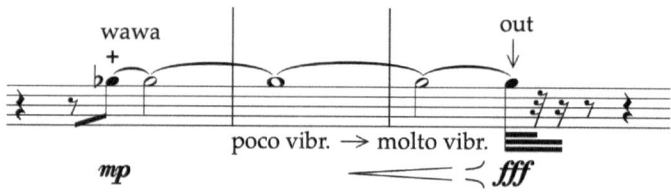

Fig. 10.6 'Sound shot' effect on the trombone

The following phase is a kind of reminiscence of what has already happened, with elements of a rhythmic and intonational canon, with a retrograde of separate rhythmic structures and sounds (and their interactions) etc.

At the end of this piece, it all comes down to the repetition of one note (Fig. 10.7). The saxophone plays one note using the slap tongue technique, which sounds like a heartbeat. This acoustic effect that musically expresses the heartbeat and simultaneously embodies a development of the idea of resonance. This was done for a reason.

As already noted, each element of a given work is a response to something and the beginning of another element. For example, effects with a combination of playing ordinario and smorzato appeared throughout the piece, and the final rhythm (heartbeat) embodied the rhythmic formulas that had already been featured in the dramaturgy of this composition.

The culmination point was quiet, like some kind of transition into another dimension, where the effect of resonance was demonstrated on an acoustic level. This piece demonstrates one of the approaches to writing ensemble music.

Fig. 10.7 Last page of the score of Rakhat-Bi Abdyssagin *The Space of Resonance* (2018), heartbeat effect is performed by saxophone at the very end (© Rakhat-Bi Abdyssagin (2018). All Rights Reserved)

References

Abdyssagin, R.-B. 2018. *The Space of Resonance* for 7 Performers.

Abdyssagin, R.-B. 2020. *13 Notes from the Parallel Universe* for String Quartet.

Abdyssagin, R.-B. 2022. *Noveishie kompozitorskie i ispolnitel'skie tehniki* (The Newest Composing and Performing Techniques). Astana: KazNUA.

Aharonov, Y., P.G. Bergmann, and J.L. Lebowitz. 1964. Time Symmetry in the Quantum Process of Measurement. *Physical Review* 134 (6B): B1410-1416.

Aharonov, Y., and L. Vaidman. 1998. On the Two-State Vector Reformulation of Quantum Mechanics. *Physica Scripta* T76: 85–92.

Bach, J.S. 1751. *Die Kunst der Fuge*, BWV 1080. Berlin: C.P.E.Bach.

Birkhoff, G., and J. Von Neumann. 1936. The Logic of Quantum Mechanics. *Annals of Mathematics* 37 (4): 823–843.

Brukner, C., and A. Zeilinger. 2001. Conceptual Inadequacy of the Shannon Information in Quantum Measurements. *Physical Review A* 63 (2): 022113.

Covey, S. 2013. *The 7 Habits of Highly Effective People*. 25th Anniversary Edition. RosettaBooks LLC.

Crease, R.P. 1993. *The Play of Nature: Experimentation as Performance*. Bloomington: Indiana University Press.

Deutsch, D. 1997. *The Fabric of Reality*. London: Penguin Books.

Deutsch, D. 2011. *The Beginning of Infinity: Explanations that Transform the World*. London: Penguin Books.

Feynman, R., R.B. Leighton, and M. Sands. 2011. *Six Easy Pieces: Essentials of Physics Explained by Its Most Brilliant Teacher*. New York: Basic Books.

Griffiths, R.B. 1984. Consistent Histories and the Interpretation of Quantum Mechanics. *Journal of Statistical Physics* 36: 219–272.

Griffiths, R.B. 2003. *Consistent Quantum Theory*. Cambridge: Cambridge University Press.

Heisenberg, W. 2000. *Physics and Philosophy*. Penguin Classics.

Nancarrow, C. 1948–1992. *Studies for Player Piano*.

Pesic, P. 2014. *Music and the Making of Modern Science*. Cambridge, MA: MIT Press.

Rae, A. 2018. *Quantum Physics: Illusion or Reality?*, 2nd ed. Cambridge: Cambridge University Press.

Rovelli, C. 1996. Relational Quantum Mechanics. *International Journal of Theoretical Physics* 35: 1637–1678.

Schottky, W. 1921a. Das Kausalproblem der Quantentheorie als eine Grundfrage der modernen Naturforschung uberhaupt. *Naturwissenschaften* 9 (25): 492–496.

Schottky, W. 1921b. Das Kausalproblem der Quantentheorie als eine Grundfrage der modernen Naturforschung uberhaupt. *Naturwissenschaften* 9 (26): 506–511.

The UIUC Talkshow. 2023. *Tony Leggett. The UIUC Talkshow #33*. https://www.youtube.com/watch?v=1JrqtZxmT40 (Accessed 9 May 2023).

't Hooft, G. 2016. *The Cellular Automaton Interpretation of Quantum Mechanics*. Springer Science+Business Media.

Vedral, V. 2023. The Everything-Is-a-Quantum-Wave Interpretation of Quantum Physics. *Quantum Reports* 5: 475–480.

Whittall, A. 2008. *The Cambridge Introduction to Serialism. Cambridge Introductions to Music*. Cambridge and New York: Cambridge University Press.

11

On Time in Music

Music is a unique art. In a sense that it does not exist without time. So the presence of time is essential for the existence of music. Music is not static, it is a living phenomenon that is being created each time again and again.

Almost all other kinds of art are static—they do not need time (or a process in time) to exist, and once they are created, they remain in this way. For example, a painting is a static object, and once an artist painted it, the painting does not require any further time to be born. Painting is an object and it is static, there is no time in it, it does not exist in time, there is simply no such dimension as time. The same is true for architecture, sculpture, monuments, buildings and so on. Of course, buildings may experience natural wear and tear or other damage over time or for other reasons, and the conditions of paintings may deteriorate, but this is not relevant for our discussion.

The dimension of time is not an integral part of any kind of visual art. Visual arts in general do not require time for existence. Even literature exists without any dimension of time. There is simply no element or aspect of performance in visual arts or literature. And performance requires time. It can be said that theatre and stage works as well as dance require time, and that is true, but in most cases that type of art can be unified with music in the form of operas and ballets. It is also evident that many dance and theatre performances need music.

Considering literature, yes, there is a certain element of time, but not in the same way as in music. There is no pre-defined time (or pre-defined process in time) in literature. For example, it would make almost no difference if you read a line of text in one minute or in one hour, or do you read the same book

R.-B. Abdyssagin, *Quantum Mechanics and Avant-Garde Music*,
https://doi.org/10.1007/978-3-031-63161-0_11

in one day or throughout several years. This kind of thing does not alter or change the nature of the book, while in music the time is an integral part of the music, of the art of music and performance.

A unique aspect of music is that it cannot be destroyed. Nothing can be done to music. The carriers of musical information can be demolished or annihilated, but not the music itself. If the canvas or marble or stone or paper etc. disappears/deteriorates/distorts, the same happens with painting/sculpture/architecture/literature, because visual arts and literature are embodied in these objects. However, in the case of music the score or the recording are not the music itself (this has been demonstrated above). Thus, music is truly transcendental.

Speaking about music, time is its fundamental dimension. Music exists in time. There is no music without time. Music also does not exist without space—there is no music in vacuum. Therefore, we can say that music exists in space–time. But what is a time? And what is a space?

In the theory of relativity Albert Einstein displayed 'the equivalence of matter and energy' (Raine 2014, p. xiii). As Einstein himself wrote: 'Physical objects are not in *space*, but these objects are *spatially extended*. In this way the concept "empty space" loses its meaning' (2014) and 'According to the general theory of relativity, the geometrical properties of space are not independent, but they are determined by matter' (2014, p. 115).

Another fundamental facet of the theory of relativity is that it unifies space and time (2014, pp. 56–57). This shows that space and time is a single phenomenon and there cannot be space without time or time without space. They do not exist without each other. As Paul Davies wrote: 'Indeed, the central lesson of the theory of relativity is that space and time are not merely the arena in which the drama of the universe is acted out but part of the cast. That is, space–time is as much a part of the physical universe as matter; in fact, the two are intimately interwoven' (1989, pp. viii–ix). This quotation refers to Einstein's 1915 general theory of relativity, where space–time geometry is dynamically linked to matter.

Hermann Minkowski said: 'Space of itself, and time of itself will sink into mere shadows, and only a kind of union between them shall survive' (Feynman et al. 2011, p. 109). This Minkowski quote relates to Einstein's 1905 special theory of relativity, which links space and time themselves.

As formulated by Stephen Hawking: 'In 1915 Einstein introduced his revolutionary general theory of relativity. In this, space and time were no longer absolute, no longer a fixed background to events. Instead, they were dynamical quantities that were shaped by the matter and energy in the universe. They were defined only within the universe, so it made no sense to talk of a time before the universe began' (2022, p. 8).

St Augustine of Hippo, who was born in 354 and died in 430, provides exceptionally deep and sometimes shockingly modern vision and interpretation of the nature of time in his *Confessions*, Book XI 'Time and Eternity'. His ideas about time are absolutely fascinating, especially if taking into account that they were written by a man who lived in the fourth and fifth centuries, sixteen centuries ago. St Augustine writes that 'there was no 'then' when there was no time'. [...] Nor was there any time when time did not exist' (2008, p. 230). Concerning the passage of time, St Augustine examines the ontology of past and future tenses by saying 'how can they 'be' when the past is not now present and the future is not yet present? Yet if the present were always present, it would not pass into the past: it would not be time but eternity. If then, in order to be time at all, the present is so made that it passes into the past, how can we say that this present also 'is'? The cause of its being is that it will cease to be. So indeed we cannot truly say that time exists except in the sense that it tends towards non-existence' (2008, p. 231). St Augustine argues about the existence of past only in the minds of people, in memory and perception, and finally concludes that 'it is in you, my mind, that I measure periods of time' (2008, p. 242).

Heisenberg referred to *Confessions* while writing about the nature and mystery of time, and paraphrased St Augustine's thoughts: 'Only for us is time passing by; it is expected by us as future; it passes by as the present moment and is remembered by us as past. But God is not in time; a thousand years are for Him as one day, and one day as a thousand years. Time has been created together with the world, it belongs to the world, therefore time did not exist before the universe existed. For God the whole course of the universe is given at once. There was no time before He created the world' (2000, pp. 78–79).

Heisenberg highlights some crucial differences between ancient thought (represented by St Augustine) and the attitude of modern science. In particular he discussed the meaning, potential applicability and limitations of any word and concept behind it by saying that 'the word 'created' at once raises all the essential difficulties. This word as it is usually understood means that something has come into being that has not been before, and in this sense it presupposes the concept of time. Therefore, it is impossible to define in rational terms what could be meant by the phrase 'time has been created.' This fact reminds us again of the often discussed lesson that has been learned from modern physics: that every word or concept, clear as it may seem to be, has only a limited range of applicability' (2000, p. 79).

This elucidates that there was no time before the universe (space), and there may not be time after the universe (space). Time is a specific dimension that was created with the universe and may disappear with the universe. Therefore, the question of 'what was before the universe' is meaningless, as there was no 'before' before the universe, and there might be no 'after' after the universe. Continuing the idea of St Augustine paraphrased by Heisenberg, it can be said that time is an illusion; for God time does not exist. There is no before or now or after, past or present or future, as for the Creator the time itself is given at once.

Analogous to this, each piece of music is born with its own time. Each musical composition is a universe of its own. There was no 'this composition's time' before 'this composition', and there will be no after it. And a composer takes the role of a creator, for him the entire course of his whole work is given at once. A composer knows simultaneously all the past, present and future within the process itself. A composer 'creates' the time within his music composition, and perceives the time and 'tenses' of his composition all at once.

It can be argued that while single notes or passages within the composition do not know themselves their path, past and future, and everything seems as undetermined for them, for the composer knows all. Single notes and elements within musical composition do not know their role, or what was before or what will be after them, but the composer knows it. Metaphorically, the following philosophy can be deduced: people, like single notes within the music, do not know what was (precisely) before them or what will be after them, and they live within the time of the universe, while God knows all of His universe, like composer knows all the composition. So, in this case the composer assumes the role of a creator of music. And each composer's musical piece is a universe of this composer created by this composer.

My symphonic picture *Time Run* for orchestra (2018) is dedicated to the philosophy of time (Fig. 11.1). The epigraph of this my symphonic work is:

> People say: time passes.
> Time says: people pass.

The world premiere of this orchestral work was held in the Grand Hall of the D.D. Shostakovich Saint Petersburg Philharmonia by 'Taurida' Symphonic Orchestra of Saint Petersburg on a final concert/closing ceremony of the II International Tchaikovsky Festival for Young Musicians, 31st May 2018, Russia.

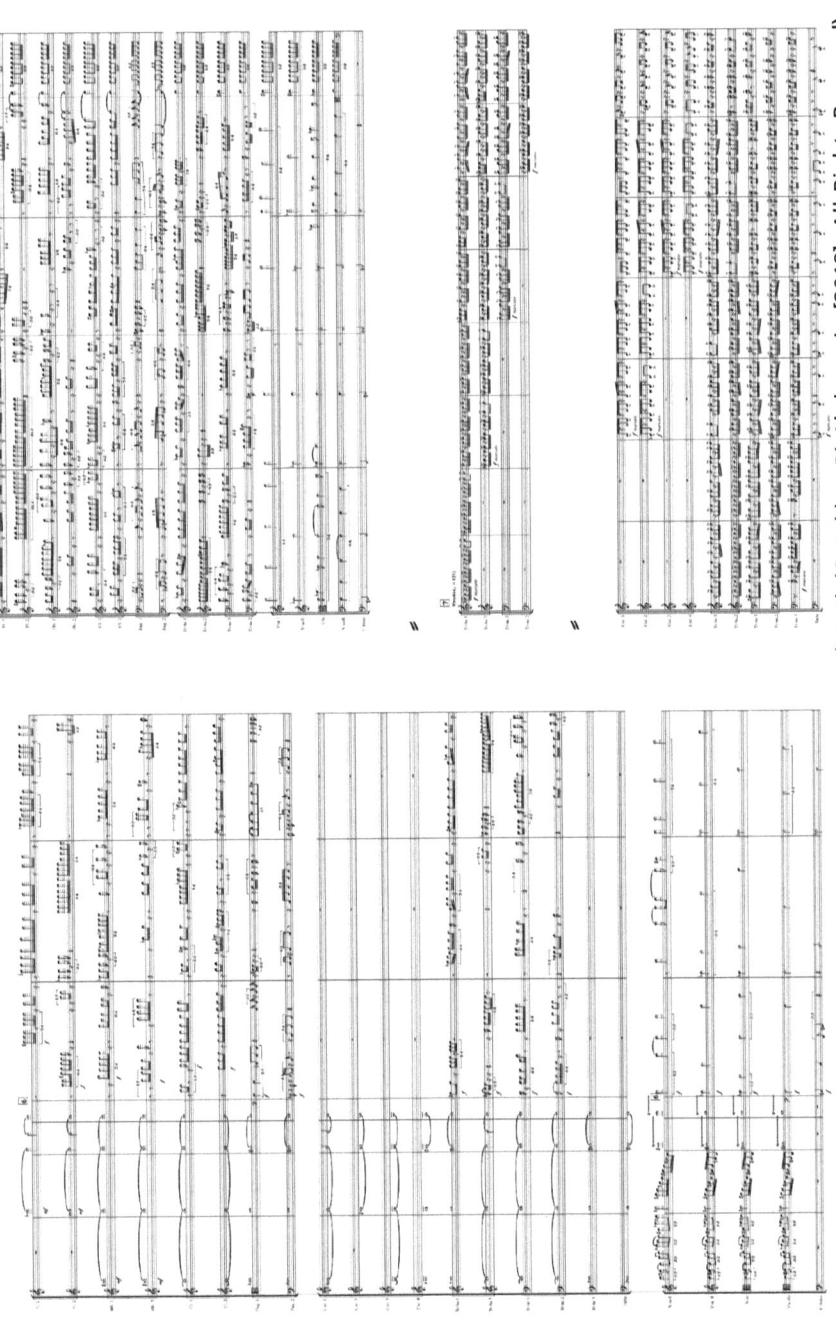

Fig. 11.1 Excerpt from the score of Rakhat-Bi Abdyssagin *Time Run* (2018)

Indeed, time is one of the most enigmatic and mysterious phenomena. There have been numerous competing theories of time. For example, physicist Julian Barbour for a long time has been developing a completely new theory of time that implies a completely new picture of reality (1999, 2020). Barbour and his collaborator Tim Koslowski have developed a theory of experienced time in a fundamentally timeless world (as Barbour wrote during our correspondence in November 2023).

And as Richard Feynman once said jokingly, 'Time is what happens when nothing else does' (Barbour 1999, p. 2).

Here I conduct structural analysis (2022, pp. 92–96) of my composition *Serenata di stelle invisibili* (Serenade of Invisible Stars) for flute, clarinet, alto sax, cello and piano (2019). Shortly about the inspiration and how the title of this piece emerged. As well-known, the speed of light is 299,792,458 m per second, commonly approximated as 300,000 kms per second. In around 8 min light travels from Sun to Earth. From one end of a galaxy to another light can travel millions of years (if not more). And a galaxy—in the present-day observable universe—is incomparably less than a grain of sand. When we look to the stars, we do not see the stars themselves, but only light that comes from them. And light from distant stars reaches us in millions of years. So what we see now is the condition of the star in which it was millions of years ago. It may not exist now. Even that specific celestial combination may not exist now… But we see it because light from it comes only now (2013, pp. 12–13). That's why the composition is named *Serenade of Invisible Stars*. The epigraph to this piece is: when we look into the sky, we think that we look into the future. But in reality we observe the past.

This conceptual idea also unites *Serenata di stelle invisibili* with another piece of mine—*Petrogliff* for two pianos (2011). However, *Serenata di stelle invisibili* is realised using a completely different notation system and timbral-textural structures, both on the level of syntax and semantics.

Serenata di stelle invisibili was composed in 2019 as my final work for the Master di II Livello (analogue of artistic doctorate) in composition at Conservatorio di Musica Giuseppe Verdi di Milano. The world premiere took place on 25 October 2019 in Sala Puccini, performed by the Syntax ensemble, artistic director Maurilio Cacciatore, conductor Pasquale Corrado. It was symbolic for me that this piece was performed at the same concert with the works of Beat Furrer.

Immediately, from the first page of the score, It is evident that the traditional method of notating rhythm (fixing time in the usual durations, bars) was rejected in favor of a more covariant (in this particular case) approach of notation using seconds, in particular, an 'end-to-end' system of second-by-second notation is used (Fig. 11.2). Starting from 0'00", 1", 2", 3", 4" and so on until the end of the piece. The seconds are indicated at the top of the score, and from each second there is an arrow down, giving a time (rhythmic) designation. The specific second (the moment of entry) is indicated only if a new sound event occurs (the entry of an instrument, the beginning or end of playing, etc.). If no event occurs, then the corresponding second is skipped and the next one is indicated, the one that correlates with the entry (or completion) of a new element.

At 0'00" the pianist plays A_4. At 1" the pianist removes the pedal (continuing to hold A_4). The clarinet enters at 2", and the flute and cello enter at the same time at 3". 4"– saxophone introduction, 5"– changes in the cello part, 6"– the flute part changes and the pianist removes his right hand from the note A_4 and takes the cluster with his left hand.

On the first page of the score, sound events occur every second, so all consecutive seconds are indicated (in this example, from 0'00" to 20"). On the remaining pages it can be seen that not all seconds are displayed, but only those when something new is introduced.

By analysing the 'tension' and 'density' (as well as its changes) in the appearance of seconds, one can already obtain a certain idea of the development of the form and dramaturgy of the composition.

The performers play not from the individual parts, but from the full score. There are different techniques for playing similar scores notated second-by-second. One of the common ones is that each instrumentalist has a stopwatch in front of them, and they start them synchronously. At the world premiere in Milan, conductor Pasquale Corrado showed the performers 5 fingers of one hand, counting 5s, then counting 5s again, and repeating this cycle until the end of the piece. This was accompanied by the conductor's instructions to the performers about the moment of their entry.

Let us carry out pitch and sonoristic analysis. In particular, about the formation of a 'campo armonico' (the harmonic field) of this composition. The first sound of the piece is the note A_4 played on the piano. Then the clarinet plays B♭, and given that the score is transposed, the sound result is A♭. Then the flute plays E—the third harmonic in the scale from the sound A. The cello takes A quarter sharp and moves to A regular. The saxophone plays G♯ (written), which will sound as H (because of transposition). Consequently, at the beginning of the first page there is a general 'singing' around the tone A, which for a while became the center of 'gravity' in the fabric of music. The entire pitch/frequencies co-ordinate of the composition is built according to a similar principle. And this process operates not only at the

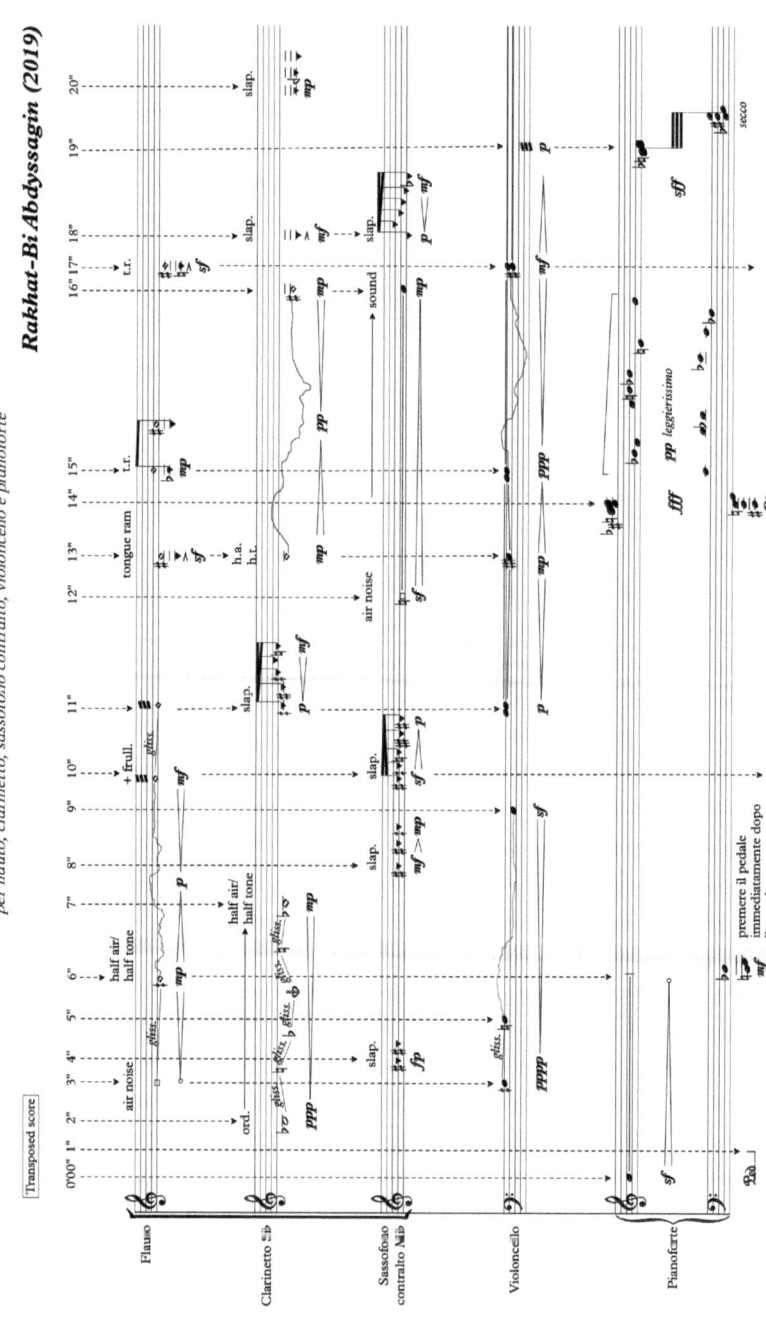

'strategic' (global), but also at the 'tactical' (local) level. For example, the flute part from 3" to 11" is a 'spiral' movement around the tone E. The saxophone at 4", 8" and 10" seconds plays descending 'impulses' based on the closest possible frequencies.

Microtones are used in abundance in this piece. On the one hand, they have a function of a controlled glissando, on the other hand, they appear as separate elements of the sound series. Many 'harmonic' solutions in a work are determined by a tone row (series), but of 24 tones, rather than 12. That is why there was a need for the active use of microtones.

The principles for determining the main 'tones' of a work are in a certain way similar to methods of game theory and probability theory, as well as to the experiments of John Cage. The work with micro-intervals has common ground with the way Luigi Nono operated with micro-intervals in his orchestral work *A Carlo Scarpa, architetto, ai suoi infiniti possibili*.

Now I make sonoristic timbral-textural analysis of the first page. The piano creates a peculiar acoustic effect when the pedal is removed and he note A is held. Afterwards the clarinet plays a glissando, gradually moving from ord. to half air / half tone. The flute starts playing with air noise, gradually moves to half air / half tone, to which frullato is then added. At the same time, the cello plays a microtonal glissando, and the saxophone sharply plays the slap tongue technique. The pianist also uses the 'delayed' pedal technique to 'capture' not the sounds themselves, but their echoes.

The slap tongue of the saxophone grows by the method of 'accumulation', combines both percussive and glissando-like elements in its implementation, and is transferred to the clarinet through a 'chain reaction'. The tongue ram on the flute sounds as an 'echo'. The glissando-like structures continue to sound and are taken to a new level when the cello plays one note on two strings (one of them open). Gradually, one note changes through the glissando, while the other continues to sound unchanged. An obvious sonorous polyphony appears.

On the second page the original line of development continues—the combinatorial nature of air noise, glissando, slap tongue and other techniques. Bartók pizzicato in the cello part (41") marks dramatic phase changes. On the one hand, a 'shot' effect is achieved, similar to a beam of energy (quantum), on the other, because of the interplay of two notes (D on the open string and C♯ on the G string), a 'bridge' is created between the cluster-like sonorities of the piano and slap in wind instruments.

Homogeneous and similar in sonorous nature, extended performing techniques layer the entire texture of the work for all 5 instruments, creating a single sound space. At the same time, each instrument leads its own 'quasi-independent' and autonomous line, develops it and brings it to a new sound dimension (Fig. 11.3).

Multiphonics appear one after another, first in the clarinet part, then performed by flute, and pass through all stages of sonoristic culmination. Multiphonics sound in all wind instruments, combining air noise, smorzato, with varying dynamics. Each of the multiphonics has its own 'breath', a flash time and a moment of attenuation, and is reborn again in a different quality. The cello echoes these sound phenomena and uses a rare technique: playing one sound (on different strings) simultaneously pizz. and arco (1'19"). In the course of further development, aleatoric methods were also used.

One of the culminating episodes (not only in dynamics, but also in meaning) is page number 10, a tremolo between the multiphonics of flute and clarinet (Fig. 11.4). This creates correlation with the bisbigliando technique that came before. The cello begins its part by playing without a bow (only with the fingers). And during the tremolo of multiphonics in the wind instruments, a new technique is used by cello: changing the pressure of the bow on the string creates unique and diverse harmonic clusters and combinations. This and similar techniques can be called as cello multiphonics.

At the very end of the piece (Fig. 11.5), the element of resonance is used. A series of pauses that appear throughout the score are very important and serve as a transition to another 'space', from which new music is born, just as new worlds emerge from the void (Fig. 11.6).

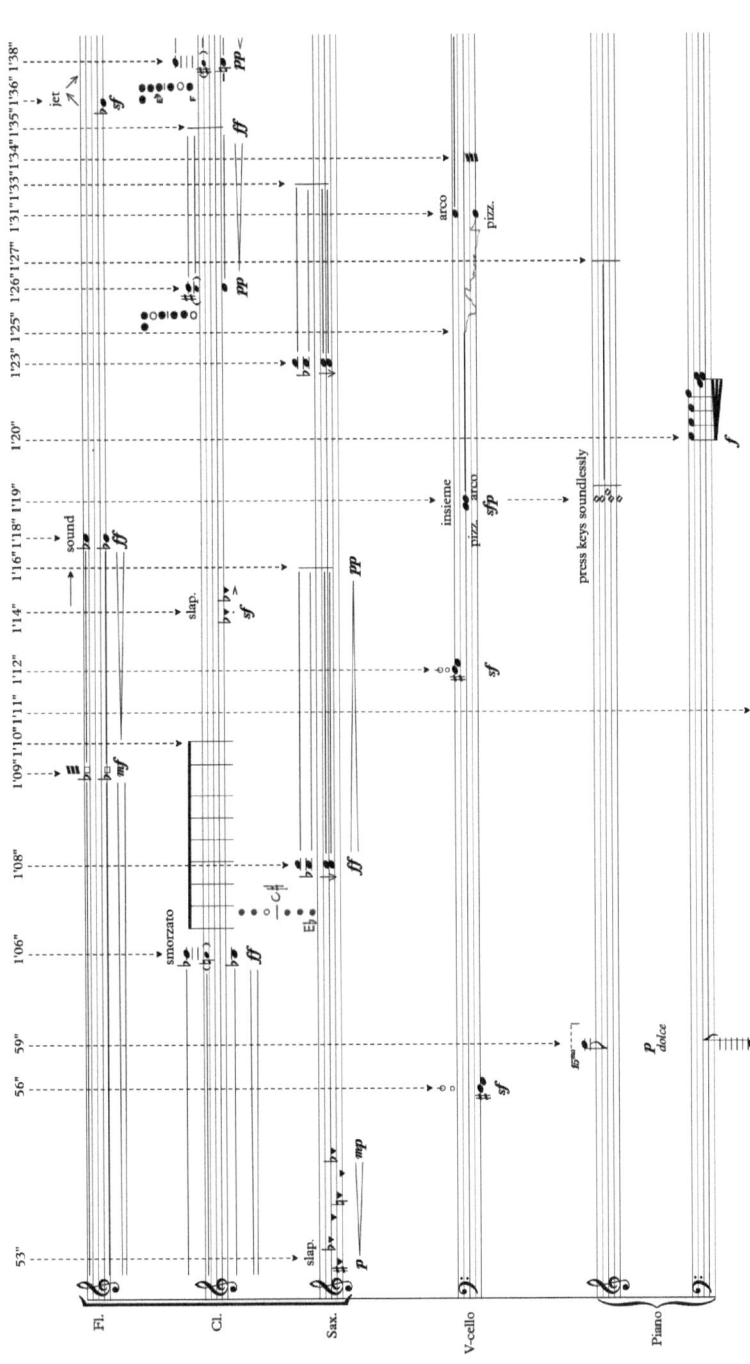

Fig. 11.3 Beginning of the intensification phase, excerpt from the score of Rakhat-Bi Abdyssagin *Serenata di stelle invisibili* (2019) (©
Rakhat-Bi Abdyssagin [2019]. All Rights Reserved)

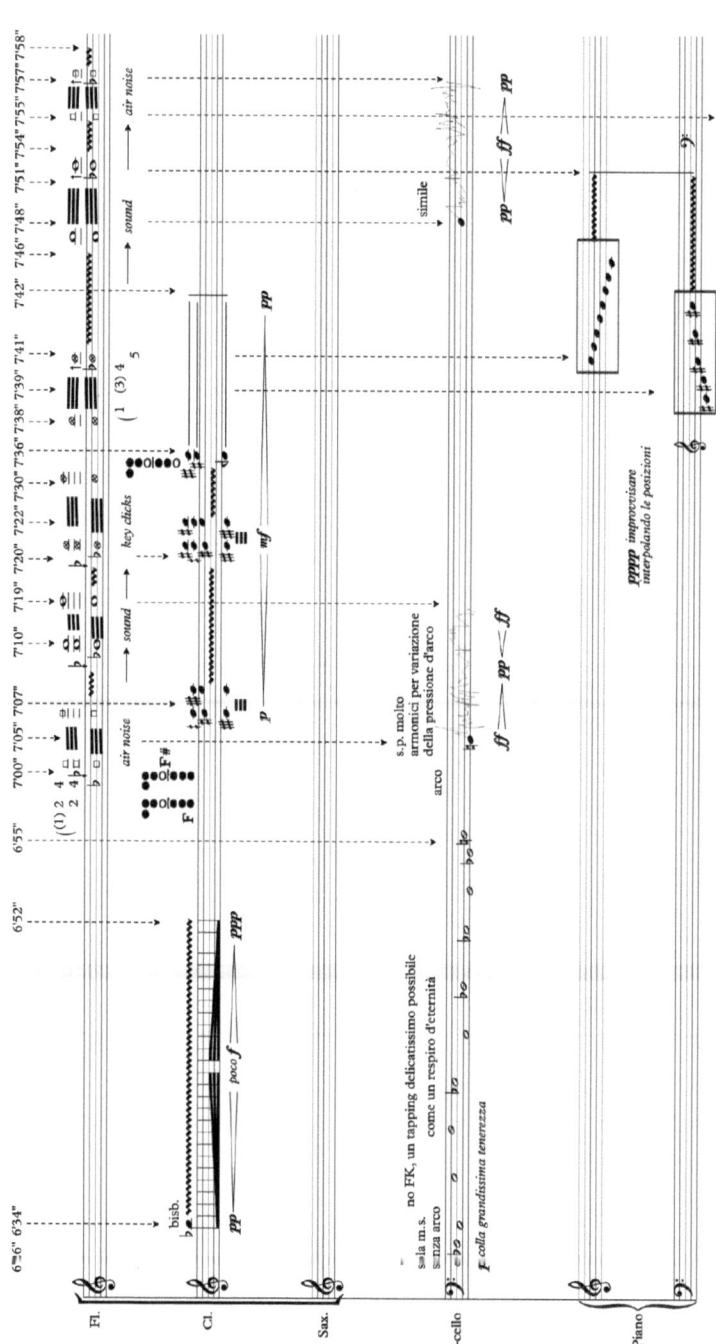

Fig. 11.4 Culmination phase, page 10, excerpt from the score of Rakhat-Bi Abdyssagin *Serenata di stelle invisibili* (2019) © Rakhat-Bi Abdyssagin [2019]. All Rights Reserved)

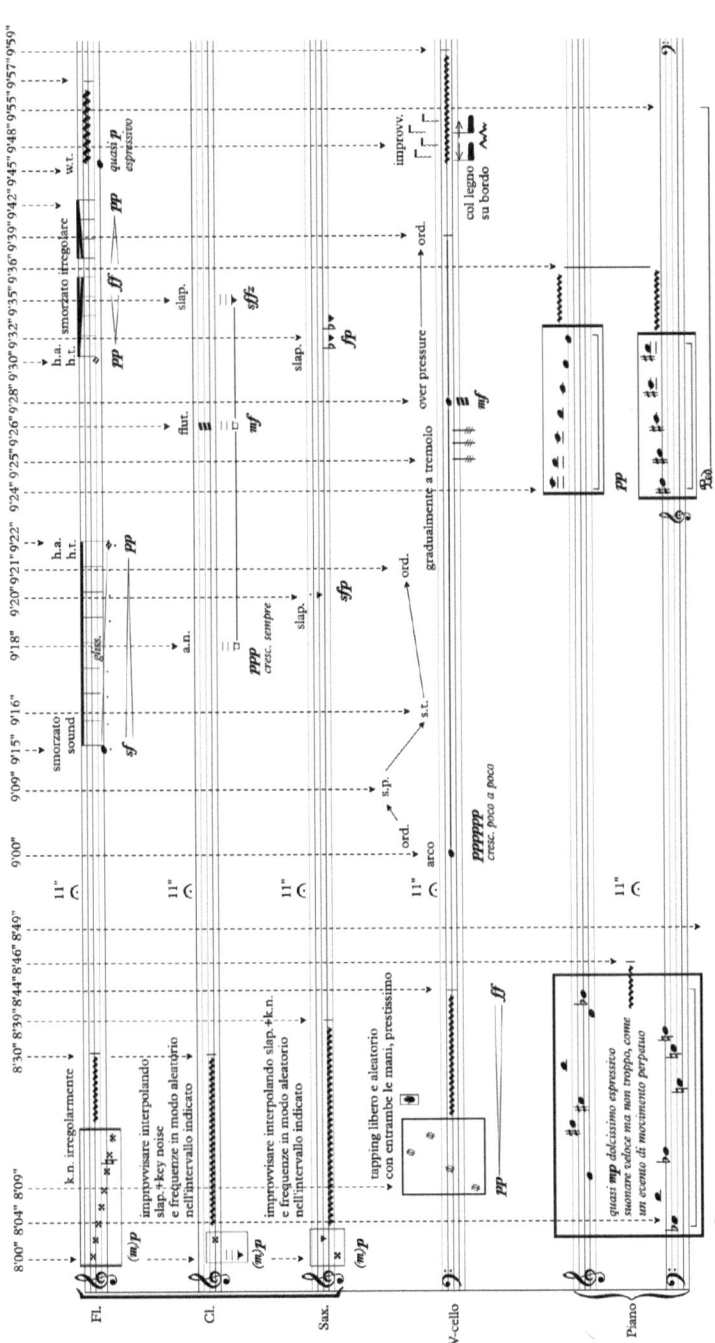

Fig. 11.5 Post-culmination phase, excerpt from the score of Rakhat-Bi Abdyssagin *Serenata di stelle invisibili* (2019) (© Rakhat-Bi Abdyssagin [2019]. All Rights Reserved)

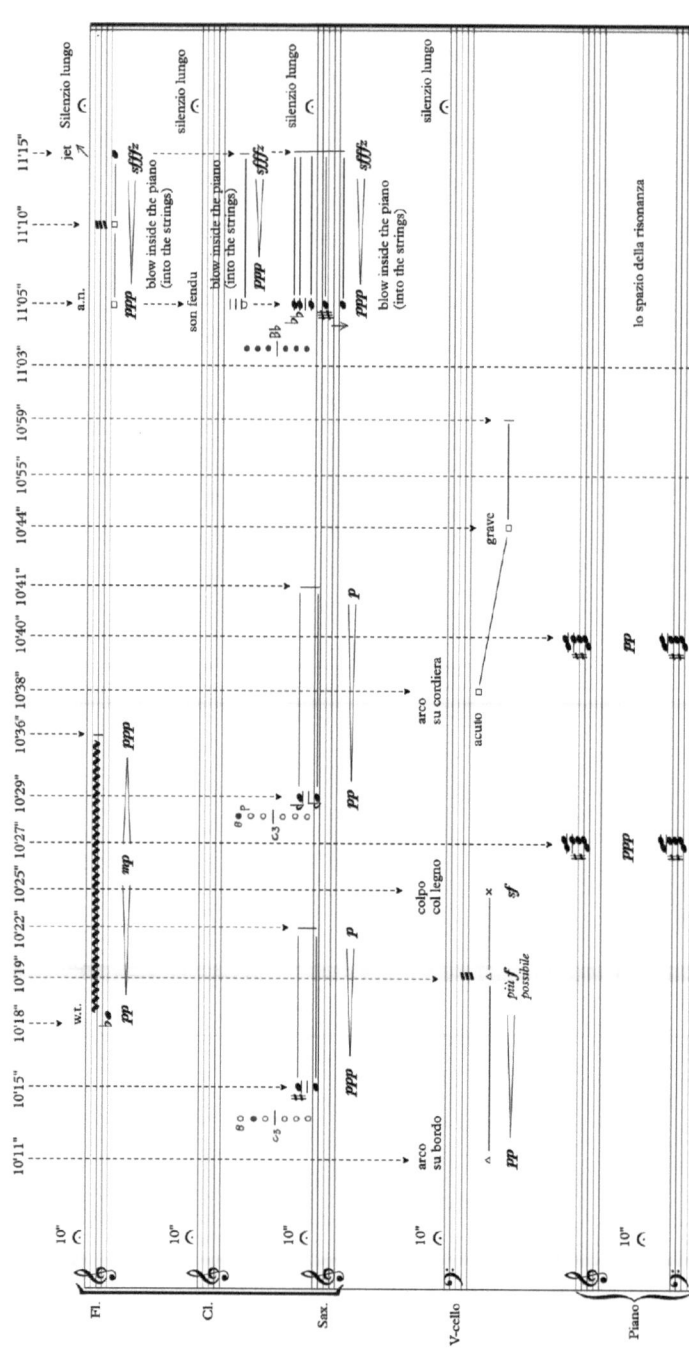

Fig. 11.6 Final page of the score of Rakhat-Bi Abdyssagin *Serenata di stelle invisibili* (2019) (© Rakhat-Bi Abdyssagin [2019]. All Rights Reserved)

References

Abdyssagin, R.-B. 2013. *Mathematics and Contemporary Music*. Almaty: Kazak Universiteti.

Abdyssagin, R.-B. 2018. *Time Run* for Symphony Orchestra.

Abdyssagin, R.-B. 2019. *Serenata di stelle invisibili* per flauto, clarinetto, sassofono contralto, violoncello e pianoforte.

Abdyssagin, R.-B. 2022. *Noveishie kompozitorskie i ispolnitel'skie tehniki* (The Newest Composing and Performing Techniques). Astana: KazNUA.

Augustine, St. 2008. *Confessions*. Translated with an Introduction and Notes by H. Chadwick. Oxford World's Classics. New York: Oxford University Press.

Barbour, J. 1999. *The End of Time: The Next Revolution in Our Understanding of the Universe*. London: Phoenix, Orion Books, an Hachette UK company.

Barbour, J. 2020. *The Janus Point: A New Theory of Time*. Bodley Head.

Davies, P. 1989. Introduction. In *Physics and Philosophy*, W. Heisenberg, vii–xvii. Penguin Classics.

Einstein, A. 2014. *Relativity*. London and New York: Routledge Great Minds.

Feynman, R., R.B. Leighton, and M. Sands. 2011. *Six Not-So-Easy Pieces: Einstein's Relativity, Symmetry, and Space-Time*. New York: Basic Books.

Hawking, S. 2022. *How Did It All begin? Brief Answers, Big Questions*. London: John Murray Publishers.

Heisenberg, W. 2000. *Physics and Philosophy*. Penguin Classics.

Nono, L. 1984. *A Carlo Scarpa, architetto, ai suoi infiniti possibili*.

Raine, D. 2014. Introduction. In *Relativity*, A. Einstein, xi–xvii London and New York: Routledge Great Minds.

12

Symbols, Mathematics and Notation

As already quoted in Chapter 10, 'we all know that "the map is not the territory." A map is simply an explanation of certain aspects of the territory" (Covey 2013). So the score is not the music itself, but only a description of the music. The same applies to recordings (or devices on which they are encoded) and their files which are also not the music itself. 'Consider for example a piece of music. What is it? It is certainly not the paper and ink used to write out a copy of the score, neither is it the compact disc on which a particular performance is recorded. It isn't even the pattern on the disc or the vibrations in the air when the music is played' (Rae 2018, p. 64).

The notation is not the music, it is only a symbolic language invented to 'encode' and 'capture' the musical ideas. Of course, in a number of cases notation and its possibilities as well as limitations may affect the music, and the ability to read/interpret notation strongly influences the performing aspects of music. However, recognising all its importance, notation remains a language of symbols, a system for transmitting information about music from a composer/creator to performers. It is possible to suggest that the emergence of crucial elements of notation shares some similarities with the emergence of the fundamental concept of number and, hence, the whole of arithmetic and mathematics, as described by Gottlob Frege in *The Foundations of Arithmetic* (1980).

This is another fundamental difference between music and other arts, such as literature, visual arts, painting, graphics, theatre plays and so on. In the literature the book itself is a piece of literature, in visual arts the painting itself is a piece of art, in architecture the building itself is a piece if architecture, in sculpture the monument is a piece of sculpture. However, in music a score, performance, recording etc. altogether embody music, but simultaneously are not the music itself. Music's transcendental nature led to the creation of the symbolic language of notation.

In fact, visual art usually uses 'human created' ways of depicting reality (be it objective or subjective), architecture has specific 'applied' means as places (or spaces) with the purpose of being used by people. Music, in contrast, originally has an abstract nature and is 'natural' in the sense that music already exists in nature in the form of sounds and silence, while composers, instrumentalists and other musicians are only 'shaping' the phenomena that exist in nature. Yes, some may perceive that painting depicts reality, or that objects being depicted exist in reality, but the painting itself is never the reality itself that it depicts. Sounds and silence always remain in reality, and composers 'give form' to this reality in the sense of organising (speaking poetically) the space and time of music, as well as two basic states (bits of information) of music—silence and sound. Just as a retrospective, displaying such natural forces of music has been one of the philosophical ideas behind John Cage's renowned 4′33″ according to Kyle Gann (2010).

In order to further elucidate the relationship between music and score, let's consider the concepts of *types* and *tokens*. As Christopher G. Timpson wrote, 'proposition—what is said—and sentence—what is used to say it—are different things' (2013, p. 17). In a similar way music—the information being transmitted/encoded—and the score—carrier/encoder of this information—are different entities. There is a clear difference between the information itself and the carrier of the information, just as there is between a thought/idea and the words/sentences used to encode that thought/idea.

As Timpson continues, 'What is, on each of these occasions, produced (a particular pattern of soundwaves corresponding to the phonemes making up the sentence), is called a *token* of the sentence. What each of these tokens is an instance of—what is repeatable—is the sentence *type*. Inscriptions of sentences, as well as utterances, count as tokens of particular sentence types [...] Sentence tokens are concrete things; they take up space and they exist over time. [...] Propositions and sentence types, by contrast, are *abstracta*' (2013, pp. 17–18).

In short (with a slight simplification), an information—is a type, while a specific expression of this information (for example through words)—is a token. Analogous to this, a piece of music itself—is a type, while the score of this music—is a token. One of the most wonderful and amazing facets of music is that it has many (simultaneously co-existing) tokens! The music—is a type, while its tokens are many: a score, a performance, a recording, CD/

DVD, radio, broadcasting, streaming etc. So the music is a foremost type that enjoys the power of being transmitted through a great many of tokens.

Returning to the thought that notation (and the score in general) is a symbolic representation of music, it can be metaphorically linked with the role of mathematics. I would say that mathematics is a language in which all sciences speak (2017, p. 30). Mathematics is the first and most perfect virtual reality which was created by humanity in order to describe objective reality.

Mathematics, like the notation of an aesthetically compelling score, can have its own elegance and beauty. However, while scientists can seek beauty in their theories, they 'do not take the beauty of a theory as convincing evidence for its truth' (Weinberg 2016, p. 14). The same can be said about music, though aesthetic beauty of a score can be an asset, it cannot just by itself serve as evidence of the quality of the music.

One of the best examples of describing mathematics as a language used by natural sciences was provided by Werner Heisenberg (2000): 'The concepts of the general laws must in natural science be defined with complete precision, and this can be achieved only by means of mathematical abstraction [...] In theoretical physics we try to understand groups of phenomena by introducing mathematical symbols that can be correlated with facts, namely, with the results of measurements. For the symbols we use names that visualize their correlation with the measurement. Thus the symbols are attached to the language. Then the symbols are interconnected by a rigorous system of definitions and axioms, and finally the natural laws are expressed as equations between the symbols' (p. 117).

What is mathematics? Is it a human invention? Or do we discover it like discovering laws of nature? Does mathematics objectively exist in reality? Or was it here before humanity? These are among the foremost foundational questions of the philosophy of mathematics.

I would repeat that mathematics is a language in which all sciences speak. Mathematics is the first and so far the most perfect virtual reality created by humankind in order to explain objective reality. Mathematics is both an invention and a discovery. There was no mathematics before humankind. Mathematics is the result of interaction of the human mind with the objective reality of our world. Mathematics is one of the most important phenomena which made humankind a humankind. It is a human invention, but not an ordinary one, probably one of the greatest ever made.

It is very natural for us to think that mathematics is inseparable part of the objective reality, a language in which the universe is created, and perceive it as the foremost truth that exists. Nevertheless, this view is only partly true.

'2 + 2 = 4' seems to be an unshakable and ultimate truth. But it is not. In the sense that it is not as true as the objective fundamental laws of nature. Mathematics is based on axioms. These axioms and their reflections, just like $2 + 2 = 4$, cannot be proven experimentally. The fact that mathematics is not verified by experiment separates it from natural sciences. As Richard Feynman wrote: 'Mathematics is not a science from our point of view, in the sense that it is not a *natural* science. The test of its validity is not experiment' (2011, p. 47). Indeed, you cannot experimentally prove mathematics. All result of mathematics is a work of mind, and it is humans who attribute meanings to numbers and create them. As A. Rae wrote, 'the apple fell under gravity without having to do any mathematics!' (2018, p. 125). Wolfgang Pauli sometimes even used terms such as 'mathematical fiction', 'mathematical tricks' (1946), 'mathematical tools' and 'mathematical formalism' (1958).

$2 + 2 = 4$ is an axiom (more correct would be to say that it is a theorem), it is not the law of nature, but rather our human view of the nature. It is not possible to empirically verify and validate mathematical conjectures and theories. Mathematics is proven by mathematical proof, which is a logical deduction. Mathematical proof is a work of human mind. David Deutsch stated that proof is computation, and computation is a physical process, while adding that 'Mathematical facts—like Fermat's last theorem—aren't physical' (2018). Discussing this idea is out of scope of this chapter, but it is crucial to keep it in mind.

Mathematics has many kinds of proofs, nevertheless, all of these proofs have in common that they pretend to be a discovery. In a certain sense, proof is a discovery, but only within mathematics (or human mind).

Mathematics is fundamentally a human invention. Human minds collectively create mathematical axioms, rules, structures, systems, concepts and the whole architectonical framework of mathematics. But then, within these mathematical systems, mathematicians make discoveries of what is inside these systems. Once people created a mathematical world, it is possible to make discoveries inside it. But all these discoveries are discoveries inside the world of mathematics, not objective reality.

It is crucial to remember the distinction between the objective reality/ the world itself and the mathematical reality/mathematical world. When mathematicians make discoveries, they make them inside the mathematical world. In fact, mathematicians discover the possibilities objectively present in the mathematical language. Evolution of mathematical world is akin to the evolution of mathematical language.

But who exactly invented the mathematics? And why it has been so successful? Mathematics has been instrumental in defining, shaping, and ultimately creating our entire civilisation. The reason is that mathematics is not a simple invention, it was not invented in a single step. I would say that mathematics is one of the most perfect representations of the human mind, a mirror of sub-conscious meta-language of thought and meta-language of ideas. Mathematics unites humans, because the possibility of doing mathematics and the very basic foundational understandings of mathematics are inherently present in humankind as itself. Not a single person at a specific place, but step-by-step, the entire humanity participated in the creation of mathematics. Mathematics was born when the very idea of representing, picturing and viewing objective reality (the world) as abstract entities came into existence.

Mathematics was not really 'out there' to be discovered in objective reality, but it was (and remains) 'out there' in the human mind, in the consciousness of the whole humanity. This is the reason why mathematics has been invented by the entire community of thinking minds across the globe in extremely early times. Mathematics was invented for description of the objective reality, but was discovered with(in) the human mind, and was born as a result of interaction between the human mind and objective reality, the world.

The evolution of mathematics was a process of almost simultaneous inventions and discoveries. As itself, mathematics was invented, but right after the invention and creation of the very first mathematical systems, the discoveries were made within these systems.

The world of mathematical abstractions is quite close to the world of abstract ideas of Plato, so the view of the existence of mathematics in the world of abstract entities and the discovery in the world of abstract entities is called 'platonism'. However, (during our meeting on 24 November 2023) philosopher Timothy Williamson said that 'Plato was not a dogmatic thinker, his ideas (e.g. about the idea of a table) were quite mobile'. This might partly explain why Plato in the dialogues almost never reaches a firm conclusion; the question remains open even after it has been discussed and scrutinised from a variety of points of view.

There is an assumption that aliens would invent the same kind of mathematics as humans have. This statement is thought to enforce the idea that mathematics is a discovery in objective reality. However, it is not as simple as it seems, and was already considered to a certain extent (Nation 2003).

As Ian Stewart states (2017), our world is full of discrete objects that are very similar. So the human mind is well trained in dealing with a world full of similar discrete objects, like apples, grains, etc. For example, it was convenient for shepherds in the past to count their sheep, which are discrete objects and similar to each other. So the concept of numbers for counting (and symbolising) objects, as well as basic mathematical operations, are very natural to humans and to our world.

According to Stewart, if aliens were very similar to us (even humanoids), and lived in a world similar to our world (full of discrete objects similar to each other), then their mathematics would probably would be similar to our human mathematics. But let's imagine real aliens, whose world is completely different from our reality; in their world and their reality (their notion of reality, concepts, ways of thinking) there are no such things as discrete objects, there are no objects or things at all (in human sense). Then the hypothetical mathematics of these aliens may not resemble human math. The hypothetical math of these aliens may have no numbers at all, even the concept of a number may be absent, as well as basic arithmetic operations (addition, subtraction, multiplication, division) may be 'alien' to these aliens.

Probably their math will be consistent with human math, but it also might not. In that hypothetical case even the possibility of translation from 'alien' math to human math would become a question. Because the worlds in which we live are so different. Though it may appear as a speculation, in fact it is an interesting thought experiment which can provide alternative view and counterfactual argument to the discussion of the given topic. As Timothy Williamson commented (during our conversation on 1 December 2023), 'even the notion of human language and its structure might be 'alien' to those aliens, as human language consists of discrete objects'.

Concluding, mathematics is a human invention. It is a result of the interaction of human mind with objective reality.

The legendary theoretical physicist Steven Weinberg—winner of the 1979 Nobel Prize in physics (shared with Abdus Salam and Sheldon Glashow), one of the creators of the Standard Model of particle physics—wrote that mathematics 'is the indispensable language in which the principles of physical science are expressed. [...] But mathematics is not a natural science. Mathematics in itself, without observation, cannot tell us anything about the world. And mathematical theorems can be neither verified nor refuted by observation of the world. [...] The distinction between mathematics and science is pretty well settled. It remains mysterious to us why mathematics that is invented for reasons having nothing to do with nature often turns out to be useful in physical theories' (2016, p. 20).

Roger Penrose stated that 'As physics develops, there are mathematical formalisms that develop with it, and which are needed in order to express the new physical laws' (2011, p. xiv). Similarly, the evolution of music causes the emergence of new composing and performing techniques and the development of advanced methods of notation. The development of compositional thought causes the development of tools of notation.

On the other hand, since ancient times composers have been actively utilising somewhat 'arithmetical' or even mathematical means of developing and interplaying with musical material. The creation of a limited system of organising structure in music almost always leads to the establishment of inner 'laws' of that system. In the Middle Ages, the 'laws' of strict style polyphony and the ways composers could work within such a system could be considered a musical implementation of certain mathematical elements.

As long as existing natural phenomena (e.g. silence and sound) can be represented on paper by the means of a symbolic system (e.g. notation), it should be possible to implement mathematical methods to work with symbols within the system. In addition, the numeric expression of rhythms, durations, frequencies, pitches and their interrelation/variation in music led to the elaboration of mathematical ways of varying them. Almost every work of J.S. Bach can be analysed into mathematical relations not only between specific tones and harmony, but as a system of relations between tone rows (in a tonal sense) in vertical (harmonic) and horizontal (melodic) dimensions. This is also true for many other works of music.

Some specific methods of working with tone row in dodecaphony and serialism (and these methods trace their roots back to the Middle Ages!) appear as compositional calculations and computations. Retrograde, inversion, inversion of retrograde, retrograde of either only tones or durations, mutations of series and other techniques are basic tools for developing the tone row. In more advanced (and more 'liberated' in comparison with total/integral serialism) techniques, composers utilise methods of filtering (came from electronic music), harmoniser, interpolation, polarisation and others. These are essentially well-established and commonly used practices that aid composers in developing their musical material.

Some characteristics of the musical direction of *new complexity* can also be analysed from the point of view of mathematics. This is particularly true when considering complex rhythmic structures in the works of Brian Ferneyhough and others, where each tuplet is a system of its own, and subdivisions of tuplets, their compound ratios and combinations create peculiar mathematical inner correlations.

The French mathematician and Fields Medalist Cédric Villani described music as 'a mathematical art based on relations among frequencies. At 440 pulses per second you hear the note A, and if you double the frequency to 880 pulses you will hear an A one octave higher; each time you double it, you go up an octave. And if you triple the frequency to 1320 pulses you will go up to the fifth in the next higher octave, which is to say E. With only the factors of two and three, in other words, you can move from octave to octave and from fifth to fifth' (2020, pp. 41–42).

The representations are shown in the following chart (pages.mtu.edu, no date):

Note/Tone	Frequency (Hz)	Wavelength (cm)
A_4	440.00	78.41
$A\sharp_4/B\flat_4$	466.16	74.01
B_4	493.88	69.85
C_5	523.25	65.93
$C\sharp_5/D\flat_5$	554.37	62.23
D_5	587.33	58.74
$D\sharp_5/E\flat_5$	622.25	55.44
E_5	659.25	52.33
F_5	698.46	49.39
$F\sharp_5/G\flat_5$	739.99	46.62
G_5	783.99	44.01
$G\sharp_5/A\flat_5$	830.61	41.54
A_5	880.00	39.20

Middle C is C_4
$C_4 = 261.63$ Hz when $A_4 = 440$ Hz
The higher the frequency is, the smaller the wavelength

Again, it is important to remember that by its nature the spectrum and scale of sounds is continuous, it is not discrete per se. The discretisation was done historically and artificially, so there are not 7 or 12 notes, there are as many as can be—infinitely many possibilities! 7 tones and then 12 semitones are just those tones that Western music picked from the continuous spectrum scale according to the human ear. Other musical traditions, especially from the East, show greater use of microtones and a totally different attitude to the discretisation of tones.

In fact, the exact representation of the note A_4 in specific frequency varied depending on the musical era and constructional aspects of musical instruments. For example, nowadays in concert practice the tuning of the grand piano is based on $A_4 = 442$ Hz. Some piano tuners say that in the age of Mozart the tuning was much lower (perhaps even 430 Hz) because there was no technical/constructive possibility to make pianos that can withstand the tension of the strings when $A_4 = 442$ Hz. The baroque pitch is believed to be $A_4 = 415$ Hz, while 440 Hz already was a higher tone ($A\sharp/B\flat$).

On the other hand, there are historic pipe organ instruments where the A_4 is lower than 400 Hz, or around 415 Hz, or as high as 476,3 Hz or even

500 Hz. Given that the standard tuning assumes $A_4 = 440$ Hz, and half-tone higher note $A\sharp/B\flat_4 = 466,46$ Hz, a piece of music in, say, C Major would sound slightly higher than $C\sharp/D\flat$Major (in the modern system) in an organ where $A_4 = 476,3$ Hz. It is easy to imagine what will happen if the tuning of A_4 will surpass 500 Hz! The thing is that in every temperament—especially equal temperament, where the distance between two half-tones should be equal—a change of tuning in any single note will lead to a change of proportions of the whole system! Notably, even in contemporary (recently built) organs the tuning of A_4 may be different (within a certain range) in different organ stops (registers), so each stop may have its own tuning and own representation in Hz of A_4.

The organ where $A_4 = 476,3$ Hz is the prominent Hauptorgel (Große Orgel) in Dom St. Marien in Freiberg (in Saxony) built during the years 1710–1714 by Gottfried Silbermann, the legendary organ builder of the Baroque (Gottfried-Silbermann-Gesellschaft e.V., no date). On the other hand, the remarkable organ where A_4 is noticeably lower than 400 Hz is the Organ built by Louis Dubois in 1766 in Wissembourg, Église Saint-Pierre-et-Saint-Paul. Such diversity in tuning pitches can be partly explained by the fact that in the Baroque period there were many different temperaments (mostly unequal) and standards of tuning pitches (Praetorius 1619; Praetorius, 1986, pp. 30–32). In Germany, one of the main pitch standards for organ tuning was the Chorton (choir tone), where $A_4 = 465$ Hz. Another pitch standard was the Kammerton (chamber tone), where $A_4 = 415$ Hz, which was used to play together with court orchestra musicians.

Most of the Baroque organs in Germany were tuned at Chorton ($A_4 = 465$ Hz), while orchestra musicians played at Kammerton ($A_4 = 415$ Hz). This explains why in some manuscripts of J.S.Bach the organ part is written a whole tone lower compared to other instrumental parts. The reason is that in Thomaskirche Leipzig the organ (on which J.S. Bach played) was simply much higher tuned than the instruments of orchestra musicians. So the organ parts had to be transposed in order to match the tuning of other instrumentalists (Dreyfus 1987, p. 11). However, there were some exceptions to that rule. For example, Silbermann Organ in Katholische Hofkirche Dresden originally was tuned at around $A_4 = 415$ Hz, because from the beginning it was intended to be actively involved in performances with court orchestra.

Finally, in the age of baroque, the controversy and specific limitations of natural tuning led to the invention of equal temperament, which was championed by Andreas Werckmeister, who even wrote theoretical books about musical mathematics (1686) and harmony. Importantly, J.S. Bach was one of the greatest supporters of equal temperament—hence his two volumes of *The*

Well-Tempered Clavier BWV 846–893. The invention and universal accep-
tance of the equal temperament undoubtedly gave a significant impulse to
the development of music in subsequent centuries. So, the details of pitches
and tuning itself is a vast field of research in the history and evolution of
pitch standards in Western music as well as the history of changes between
different temperaments.

The above chart and discussions are centered on 12 equal temperament,
which was and mostly remains the basis of the Western European classical tradition. Nevertheless, there are many more existing tuning systems,
including but not limited to 19 equal temperament, 31 equal temperament, 53 equal temperament, 72 equal temperament and so on. Given the
continuous nature of the spectrum of frequencies, an unlimited number of
temperaments can be invented. Historically, many eastern musical traditions
are not limited to twelve-tones and have been using microtones since ancient
times, while Western European music realised and properly/systematically
discovered the microtones only at the beginning of twentieth century.

The Kazakh national instrument dombyra started to be tuned in equal
temperament only in the twentieth century, while the centuries before that it
had different temperaments and tuning pitches. If pipe organ is the musical
heart of a cathedral, dombyra is the musical heart of the Great Steppe.

Villani states that 'mathematics has become a universal language, used
everywhere in the world today, one of the very few universal languages that
currently exist. In a sense it is more universal even than music, for while
all peoples cherish music, musical conventions vary considerably from one
culture to another, whereas mathematical conventions are everywhere the
same, or very nearly so. Mathematics is a universal language in which one
knows exactly what has been stated, and in which symbols have a special
significance, not only for purposes of verification, of course, but also for
communicating very well-defined ideas' (2020, p. 22). This view can be
supported by a quote from the German mathematician David Hilbert, who
is considered one of the most important mathematicians who ever lived:
'Mathematics knows no races or geographic boundaries; for mathematics, the
cultural world is one country' (MacTutor n.d.).

This doctrine might be partially true, however, according to the perception of Feynman (2011, p. 47), Heisenberg, Weinberg and other influential scientists, mathematics is a human invention (perhaps one of the
greatest ever made). Therefore, despite its 'universality', mathematicians work
with elements existing only in human mind, while music (as sound and
silence) already exists in objective reality/nature, and composers shape natural

Fig. 12.1 D. Hilbert (*Source* Wikimedia Commons)

phenomena—sound and silence—to create music compositions. So, if mathematics can be regarded as thought compositions and thought structures, then musical works are compositions of natural phenomena.

When one of his students abandoned mathematics in order to study poetry, David Hilbert said: 'Good, he did not have enough imagination to become a mathematician' (Goodreads, no date). This illustrates the utmost importance of imagination in mathematics, because mathematics per se exists only in the human mind, it is a result of thought. Of course, Platonism as a philosophical direction believes in the existence of abstract objects (such as in mathematics), and so does mathematical realism in the philosophy of mathematics. Even music can be viewed by some as an abstract object, but its main components are natural part of objective reality.

Here we find another strong connection with physics: physicists discover natural phenomena, and composers use natural phenomena (sound and silence) to write musical compositions. Physicists describe nature using the language of mathematics (equations), while composers organise the main components of music and create composition using the language of notes and scores. And coming to 'universality', the physical phenomena of sound, silence, space and time (the main components of music) are also 'universal' (to a certain extent, of course, so we do not consider the case of a vacuum).

And yes, each culture and tradition has its own musical style and preferences, nevertheless, great music becomes the legacy of all humanity, an inseparable part of the global world, a gem in the great treasury of mankind. Both mathematics and music express and embody harmony.

References

Abdyssagin R.-B. 2017. Contemporary Music as an Echo of Time. *Bulletin of National Academy of Sciences of the Republic of Kazakhstan* vol. 1, no. 365.

Bach, J.S. 1722. *Das wohltemperierte Klavier I*, BWV 846–869.

Bach, J.S. 1740. *Das wohltemperierte Klavier II*, BWV 870–893.

Covey, S. 2013 *The 7 Habits of Highly Effective People*. 25th Anniversary Edition. RosettaBooks LLC.

Deutsch, D. 2018. *The Mathematicians' Misconception*. Transcript of a talk at the International Centre for Theoretical Physics, Trieste, Italy, on the occasion of being awarded the Dirac Medal.

Dreyfus L. 1987. *Bach's Continuo Group*. Cambridge, MA: Harvard University Press.

Feynman, R., R.B. Leighton, and M. Sands. 2011. *Six Easy Pieces: Essentials of Physics Explained by Its Most Brilliant Teacher*. New York: Basic Books.

Frege, G. 1980. *The Foundations of Arithmetic*, trans. J.L. Austin. Evanston, Illinois: Northwestern University Press.

Gann, K. 2010. *No Such Thing as Silence: John Cage's 4'33"*. New Haven: Yale University Press.

Goodreads. n.d. David Hilbert > Quotes > Quotable Quote. https://www.goodreads.com/quotes/9027059-good-he-did-not-have-enough-imagination-to-become-a (Accessed 24 February 2023).

Gottfried-Silbermann-Gesellschaft e.V. n.d. Freiberg—Dom St. Marien—Große Orgel. https://silbermann.org/orgel/freibergdom-grosse-orgel/ Accessed 23 February 2023.

Heisenberg, W. 2000. *Physics and Philosophy*. Penguin Classics.

MacTutor. n.d. Quotations. David Hilbert. https://mathshistory.st-andrews.ac.uk/Biographies/Hilbert/quotations/. Accessed 24 February 2023.

Nation, J.B. 2003. *How Aliens Do Math*.

Pages.mtu.edu. n.d. Tuning. Frequencies for equal-tempered scale, $A_4 = 440$ Hz. Accessible at: https://pages.mtu.edu/~suits/notefreqs.html (Accessed 25 February 2023).

Pauli, W. 1946. Exclusion Principle and Quantum Mechanics. *Nobel Lecture*.

Pauli, W. 1958. *Theory of Relativity*. Translated by G.Field. London, New York, Paris, Los Angeles: Pergamon Press.

Penrose, R. 2011. Introduction. In: *Six Not-So-Easy Pieces*, Richard P. Feynman. New York: Basic Books, ix-xvi.

Praetorius, M. 1619. *Syntagma Musicum II, De Organographia*. Wolfenbüttel: Michael Praetorius.

Praetorius, M. 1986. *Syntagma Musicum II, De Organographia*. Translated and edited by D.Z.Crookes. Oxford: Clarendon Press.

Rae, A. 2018. *Quantum Physics: Illusion or Reality?*, 2nd ed. Cambridge: Cambridge University Press.

Stewart, I. 2017. Xenomath! In *Humanizing Mathematics and its Philosophy*, ed. B. Sriraman, 69–80. New York: Birkhäuser.

Timpson, C.G. 2013. *Quantum Information Theory and the Foundations of Quantum Mechanics*. Oxford: Oxford University Press.

Villani, C. 2020. *Mathematics is the Poetry of Science*, trans. M. DeBevoise, Illustrations by É.Lécroart. Oxford: Oxford University Press.

Weinberg, S. 2016. *To Explain the World: The Discovery of Modern Science*. Penguin Books.

Werckmeister, A. 1686. *Musicae Mathematicae Hodegus Curiosus*. Frankfurt & Leipzig: Theodor Philipp Calvisius.

13

Heisenberg's Uncertainty Principle and Aleatoric Technique in Music

According to Werner Heisenberg's uncertainty principle, the more precisely the position of a particle is measured, the less precisely its momentum can be measured, and the other way around.

'One cannot accurately predict both the position and the speed of a particle. The more accurately the position is predicted, the less accurately you will be able to predict the speed, and vice versa' (Hawking 2022, p. 22). Heisenberg's uncertainty principle 'says that a particle's position and momentum can never be simultaneously measured' (Gubser 2010, p. 21).

If paraphrased in a wider sense, the more accurately one characteristic is measured, the less accurately the second can be measured, and vice versa.

Figuratively it can be considered that while certain parameters of aleatoric works are precisely determined, others are not declared at all. If to take frequencies and pitches as a 'position' and rhythm as a 'momentum' (speed), then certain imaginative analogies between the uncertainty principle and the aleatoric technique can be drawn.

Karlheinz Stockhausen's *Klavierstück XI* (1956) is distinguished by a poly-valent structure and a mobile form (Fig. 13.1). This piece is written in a series of separate fragments and performer is free to choose the sequence and order in which these fragments are played. So the form, dramaturgy and overall 'shape' are decided by a performer. If we consider musical form as an analogue of global 'time' in this work, then it is undetermined. However, all other parameters such as frequencies, rhythms and techniques are indicated in every single detail, thus the musical 'space' is determined in the extremely precise manner.

© The Author(s), under exclusive license to Springer Nature Switzerland AG 2024
R.-B. Abdyssagin, *Quantum Mechanics and Avant-Garde Music*,
https://doi.org/10.1007/978-3-031-63161-0_13

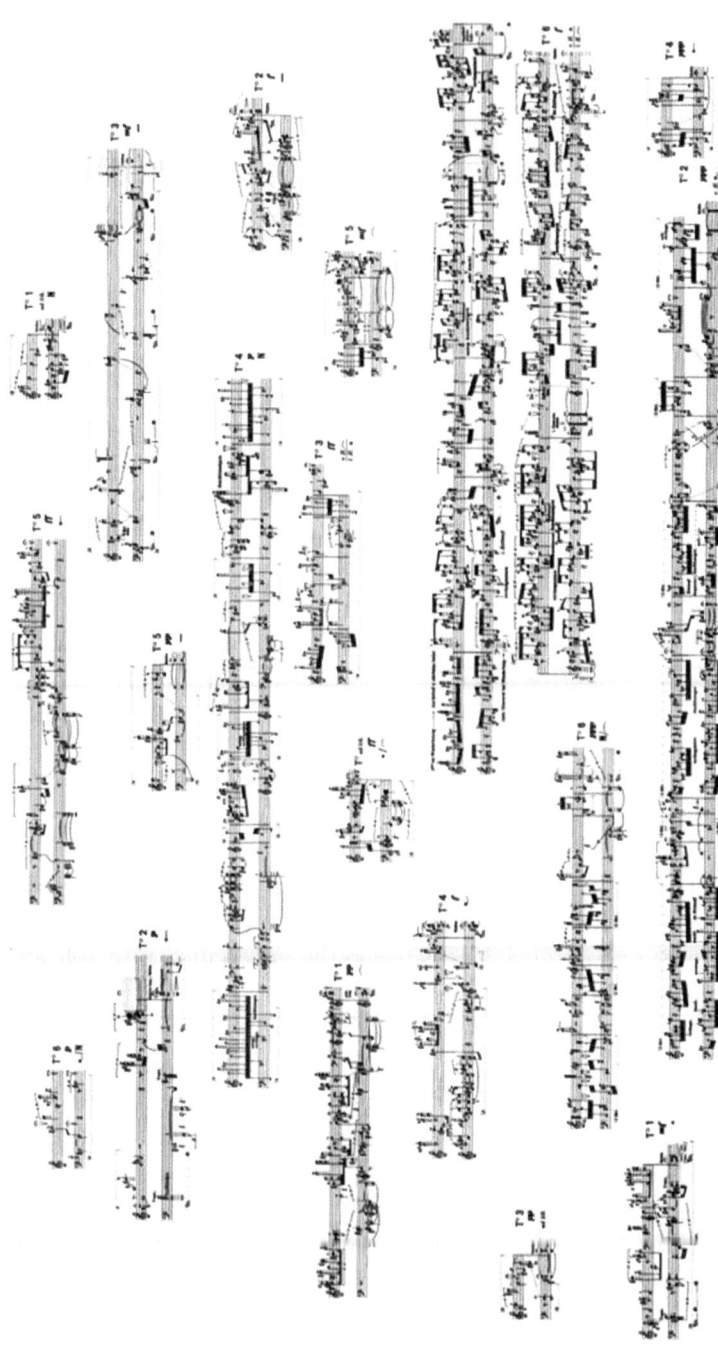

Fig. 13.1 Score of K. Stockhausen *Klavierstück XI* (© Copyright 1957 by Universal Edition (London) Ltd., London/UE12654)

This method figuratively correlates with Heisenberg's uncertainty principle. If we know the 'position' of precise elements—the combination of frequencies, rhythms, dynamics and other parameters—at a specific time, we never know their previous trajectory from the score itself. Alternatively, knowing what has already been played does not predict what will follow.

Stockhausen himself said: 'When I started to compose, after the war, there were many different directions in musical research which had been prepared by the great masters Schoenberg, Webern, Berg, Stravinsky, Bartók, Varèse. I had to go to the roots of their individual work, and find an underlying unity. It fell to me to synthesize all these different trends for the second half of the century, perhaps in a similar way that Heisenberg, in the first half of the century, had the role of bringing together the discoveries of Planck and Einstein in atomic physics' (Stockhausen 1989, p. 33). He also noted that 'We could speak of the strong influence on musicians during the early fifties, of certain books for the general reader by Einstein, or Heisenberg, of biologists like Weizsäcker, or Norbert Wiener' (Stockhausen 1989, p. 37). Considering music and mathematics, Stockhausen claimed that 'It's a universal principle that music has always been very close to mathematics, only even more complex because it must be perceived aurally' (Stockhausen 1989, p. 139). These statements show that Stockhausen was aware of many important processes in sciences, and scientific discoveries in atomic physics had a certain influence on his compositional concepts.

This is an example of Stockhausen's reflections on his own work and role in music in comparison with the work and role in physics of Heisenberg. It might even be possible to deduce the corollary of Stockhausen's phrase as a thought that Stockhausen was doing in music the same as what Heisenberg did in Physics. According to T.W. Adorno's letter to E. Steuermann from 14th October 1955, during a conversation with Adorno in Darmstadt K. Stockhausen himself claimed that 'music exhibited the uncertainty principle' (Adorno 2003, pp. 247f, cited in Adorno 2018, p. 299).

There is also another strong point of correlation with contemporary music, in particular, with aleatoric technique, open form and indeterminacy. Classical physics is usually perceived as deterministic, and in the history of science there was even a specific formulation of causal determinism known as *Laplace's demon* (though Pierre-Simon Laplace himself never called the entity a 'demon', he rather chose to refer to the entity as an 'intellect'; the word 'demon' was a later addition).

As Laplace wrote in his *A Philosophical Essay on Probabilities* (1902, p. 4) 'We ought then to regard the present state of the universe as the effect of its anterior state and as the cause of the one which is to follow. Given for one instant an intelligence which could comprehend all the forces by which nature is animated and the respective situation of the beings who compose it—an intelligence sufficiently vast to submit these data to analysis—it would embrace in the same formula the movements of the greatest bodies of the universe and those of the lightest atom; for it, nothing would be uncertain and the future, as the past, would be present to its eyes'.

Just as an additional information, not everyone agrees that classical physics is deterministic. In the article *Chaos, Quantum, Number* physicist Michael Berry wrote that 'Classical physics is not deterministic. Classical mathematics is deterministic'. Considering Laplace's demon, Prof. Berry stated, 'Assuming that mathematical determinism holds for a possibly infinite universe, and that the intelligent being is embodied—we are considering physics, after all—the future of the gas particles or billiard balls would also be sensitive to the gravity of particles in its (his? her? their?) head, or in the computers performing the calculation. Laplacian determinism ignores this self-reference, which sabotages the calculation. The conclusion is that classical physics—as distinct from classical mathematics—is not deterministic in any reasonable sense' (2023).

A thought similar to Laplace's verbal articulation of causal determinism in classical mechanics can be expressed about classical music and the tradi-tional notation. In ordinary scores (almost every existing score except open form ones) the succession of events is predefined: each note, group, phrase, measure, section etc. has its own well-defined position within the score. Looking at the score it is always possible to tell what happened before and what will happen after a certain episode. This is musical determinism that lies in the fundament of classical music and classical notation, which, as it is evident now, exerts the determinism inherent in classical mechanics in physics.

But when we come to aleatoric music, to K. Stockhausen's *Klavierstück XI*, we witness the implementation of musical indeterminism, which links greatly with the indeterminism of quantum mechanics. 'Indeterminism is linked to the fact that an object can indeed be in more than one state at any one time (e.g. you toss a coin and it can give you both heads and tails at the same time). In technical parlance this is known as a quantum superposition' (Vedral 2018a, p. 118).

Speaking of the indeterminism of quantum mechanics, it is indetermin-istic in comparison with classical mechanics and from a classical point of view, partly because quantum mechanics has no analogies in macro-world or common/ordinary reality. As Paul Davies wrote, 'unpredictability of quantum systems does not imply anarchy' (1989, p. x).

K. Stockhausen's *Klavierstück XI* is based on globally indeterministic concept. But what is particularly exciting about it? This question has fascinating reply. It was already said that in classical music, like in classical mechanics, every object has its well-defined trajectory of movement. Usually in traditional and conventional notation it is possible to know the detailed trajectory of music in the form of dramaturgical determinism. The sequence of events (notes etc.) is indicated in the score: there is only one way, a single direction between any two points in the score. One and only one way leads from A to B, from the beginning to the end. It is possible to draw only one online between any two points in the classical score. But this does not work at all in the case of *Klavierstück XI*!

The score of *Klavierstück XI* is 19 independent fragments scattered across the large sheet of paper. No one, even the composer himself, knows or controls the sequence in which these fragments are performed. Only a performer during a performance decides an order of these fragments. Yes, fragments themselves are quite detailed (except dynamics), but there is no any pre-defined order in which to perform the fragments, there is no pre-defined trajectory! 'The path of a particle through space is not generally well defined in quantum mechanics' (Davies 2011, p. xvii), and 'a quantum particle does not move along a well-defined path through space. An electron may leave place A and arrive at place B, but it is not possible to ascribe a precise trajectory linking the two' (Davies 1989, p. x).

There is no such thing as a trajectory in quantum mechanics. And there is no pre-defined or a priori known/determined trajectory of direction in *Klavierstück XI*. From this perspective it is a truly quantum–mechanical composition! Usually, for a listener, every piece heard for the first time is indeterminate, because a listener does not know what to expect, but the composer knows. This is false or illusionary indeterminacy, while *Klavierstück XI* is a true indeterminacy, as even for the composer there is no pre-defined sequence of events, and each time the work is performed it is being 'shaped' anew.

And here lies one more fundamental connection with quantum mechanics, concretely, the role of measurement. In classical physics, experiment can slightly change the object of the experiment, e.g. the measurement affects the state of what is being measured, but the degree of change is usually infinitesimal, and can be neglected. If to interpret experiment in physics as an analogy of performance in music, then there is another correlation.

In classical music (or, better say, any non-aleatoric/not-open-form) performance definitely changes the music, and this makes great performances great, and is the reason why fantastic performers provide thrilling interpretations and give new birth to well-known music. Nevertheless, the measurement, or experiment, or performance in classical music does not change the essence of music (if to perceive the form as the essence), it does not change its dramaturgy, sequence of events, and trajectory. Yes, performers may choose to play in a slightly different tempo, with a great deal of *rubato* and free interpretation, but a performer does not alter the fundamental aspects of music, first of all its position in space–time, i.e. dramaturgy, form and succession of events.

In classical music (in a very wide sense) a performer retains a certain degree of liberty of interpretation, as a rule, but is never capable of changing the essence of the work (if to consider the essence of the works as its position in spacetime, embodied in its dramaturgy, form and succession of events). During a performance, a performer may come to a specific episode in a slightly different time than another performer, but the succession of events, or what happened before and what will go after that episode, is always well-defined in the score. And speed, or tempo of performance, is also very well-known! And usually tempo (speed/momentum) is also defined in the score by the composer. And the position of any 'particle' in music as well. This is how it works in classical music (with some exceptions and extensions, which do not change, but reaffirm the rule).

However, the described situation is not the case of aleatoric music with open form, such as *Klavierstück XI*! In this work every performance changes the whole essence of the work! Just like measurement in quantum mechanics changes the reality! When performing a piece of classical music (in a wide sense), it still remains recognisable despite various aspects of interpretation. But when performing *Klavierstück XI*, it may not necessarily be recognisable as the same piece in the case of two different interpretations because every interpretation fundamentally changes the essence of the work and its reality, profoundly alters its structure and creates its dramaturgy and form anew!

'Measurements affect and change the state of the system being measured and through measurements we force the system to adopt one of its many possible states that existed prior to measurement. [...] By making use of the fact that any measurement to determine a state irrevocably changes the state we can tell when the state has been tampered with' (Vedral 2018a, pp. 122, 127). In accordance with this, in classical music measurement/performance does not heavily (beyond recognition) change the state of the system (work), while in *Klavierstück XI* it changes the whole reality of the system! And each new measurement/performance creates different states/realities of this piece. In the case of *Klavierstück XI* each new performance creates a new reality of this music, to the degree that was and remains absolutely unthinkable in Western classical music.

In any piece of classical music the position (place) and speed/tempo (momentum) are pre-defined in the score, even before any performance. So in classical music there is a certain (and detailed) reality that exists independently to measurement. But this does not happen in case of *Klavierstück XI* and quantum mechanics. In *Klavierstück XI* there is no reality beyond measurement/performance. The succession of events or their trajectory, as well as their place and speed are not *always* present; rather, they exist (materialise) only when measurement/performance occurs.

This fact can be figuratively linked with the Copenhagen interpretation of quantum mechanics, formulated by N. Bohr and W. Heisenberg. It is worth noting that though the Copenhagen Interpretation is ascribed to Bohr and Heisenberg, in fact many aspects of views of Heisenberg were totally different from the philosophical position and views of Bohr. 'The Copenhagen interpretation of quantum mechanics [...] denies that, say, an electron has a well-defined position and a well-defined momentum in the absence of an actual observation of either its position or its momentum (and both cannot yield sharp values simultaneously). [...] Thus a measurement of an electron's position creates an electron-with-a-position; a measurement of its momentum creates an electron-with-a-momentum. But neither entity can be considered already to be in existence prior to the measurement being made' (Davies 1989, p. xii).

In quite a similar way to each fragment of *Klavierstück XI* does not objectively exist in a specific place when there is no performance. Each measurement/performance creates a fragment with a specific position within the dramaturgy and timeline of the work, and each measurement/performance creates the new reality of the work, embodied in a new succession of events, and hence every new performance leads to a new form, structure and global essence of music. Performance is akin to measurement and observation, and if in classical physics observation does not radically change the reality, in quantum mechanics, as Heisenberg wrote, 'the observation plays a decisive role in the event and that the reality varies, depending upon whether we observe it or not' (2000, p. 20).

In the same way in classical music, performance does not radically change the reality of music, but in *Klavierstück XI* the reality varies depending on whether it is performed/observed or not. Moreover, each piece of classical music has its own reality embodied in a score, with all parameters—especially position (pre-defined form and fixed dramaturgy, sequence and succession of events) and momentum (speed, tempo)—well-defined, but in *Klavierstück XI* there is no such objective reality, there is no pre-defined position and momentum at all (in the sense that within each of 19 fragments, locally, there is a kind of an objective reality, but absolutely not when considering the non-local global situation of 19 fragments altogether).

The aleatoric technique and open form contain a great deal of probability, and probability is crucial to quantum mechanics. Heisenberg highlights the essential probabilistic feature by stating that 'In throwing dice we do not know the fine details of the motion of our hands which determine the fall of the dice and therefore we say that the probability for throwing a special number is just one in six. The probability wave of Bohr, Kramers, Slater, however, meant more than that; it meant a tendency for something. It was a quantitative version of the old concept of 'potentia' in Aristotelian philosophy. It introduced something standing in the middle between the idea of an event and the actual event, a strange kind of physical reality just in the middle between possibility and reality' (2000, p. 11).

Metaphorically, the form, dramaturgy and timeline of *Klavierstück XI* exist in the world of probabilities, and each new performance is like throwing a die, but with an immense number of possible outcomes. Therefore, in an imaginary way, paraphrasing Heisenberg it can be said that compositions with open form—particularly *Klavierstück XI*—exist in a strange kind of musical reality just in the middle between a possibility and the reality. Heisenberg also wrote that 'The observation itself changes the probability function discontinuously; it selects of all possible events the actual one that has taken place' (2000, p. 22). In quite a similar way in *Klavierstück XI* each new performance 'selects of all possible events the actual one that has taken place', i.e. selects among all possible combinations/structures of a form the actual one that materialised in this particular case.

Klavierstück XI is not unique among Stockhausen's oeuvres to exhibit uncertainty and indeterminacy. Some other possible implementations (both philosophical and practical) of similar approach are shown in his composition *Momente* for solo soprano, 4 choir groups and 13 instrumentalists (1962–1964, revised in 1969). Composer himself provided rigorous and profound structural analysis of this work (Stockhausen 1989, pp. 63–75). He writes about 'moments' where melodic characteristic predominates (defined in the score by a capital M), then sound qualities (K, from *Klang*), the third group based on duration (D, from *Dauer*), and finally indeterminate moments (indicated as I from *Informal*), as well as a multitude of their combinations.

Considering time, Stockhausen writes exceptionally significant observations about silence, polyphony, and superposition (another term from quantum mechanics) of independent layers:

'A third group is based on D: duration, *Dauer* in German. Moments based primarily on principles of measured durations, of different lengths, give rise to two important characteristics of any musical construction. One is silence; the other is polyphony, the superposition of more or less independent layers which are sounding at the same time. […] Silence is the result of the concept of duration: to deal with durations means to break the flow of time, and that produces silence. […] once something is cut, the pieces can not only be separated but also superimposed, since they become independent of one another. And that superimposition produces polyphony. So the principle of polyphony and the principle of silence are both based on the concept of duration and the differentiation of durations; that is why I say polyphony is the most characteristic form of articulation of moments which are based on differences of duration' (Stockhausen 1989, p. 66).

The selected works of John Cage can serve as another vivid example of the metaphoric implementation of Heisenberg's uncertainty principle in music. For example, Cage's *Freeman Etudes* (1981, 1992) is a cycle is a brilliant example of chance composition, when the whole musical layers are determined by chance operations. This is a peculiar case as in many situations the composer does not directly interfere with his immediate decisions; however, even deciding 'not to make decisions' is a decision itself. On one hand, composer precisely chooses the specific operations which should determine musical parameters, on the other hands, composer does not directly 'decide/ solve' the parameters. Moreover, in performance the tempi are to be determined by a player. And obviously, changes of tempo vastly influence the musical 'space'.

Solo for Sliding Trombone (1960)—one of the greatest pinnacles of Cage's conception of indeterminacy and uncertainty, masterfully realised in music. Here not only the rhythms, but even the pitches are not precise because Cage did not indicate the clef. A performer is even free to choose whether he will perform the whole score or only its part (and again it is up to a performer which part he will play). Additionally, trombone playing is enhanced by various extended techniques (Kasparov 2020, pp. 6–13). As Cage himself writes in performance notes to the score: 'The following 12 pages for a trombone player may be played with or without other parts for other players. It is therefore a trombone solo or a part in an ensemble, symphony, or concerto for piano with orchestra. Though there are 12 pages, any amount of them may be played (including none). Each page has 5 systems. The time-length of each system is free. Given a total performance time-length, the player may make a program (including additional silences or not) that will fill it'. Then Cage explains the difference between the size of note-heads and other aspects of performance. Even from this short introduction anyone can witness the fundamental degree of indeterminacy and uncertainty in this composition.

Interestingly, the methods of Stockhausen and Cage are opposite: Cage precisely controls the global form of the work (as a sequence of elements), but leaves a great deal of freedom for interpretation of the details, single tones and so forth. On the other hand, Stockhausen superbly details and depicts each fragment well, while their sequence and the global form are totally dependent upon performer.

V. Vedral wrote that 'quantum physics presents us with determinism at the level of the whole universe, while still implying that any subsystem in the universe is fundamentally random' (2018b, p. 91). This corresponds well with J. Cage's approach, where the global form, dramaturgy and 'shape' of the work are determined, while locally there is a great deal of freedom for a performer. Of course, mentioning Vedral's phrase, it is important to remember that it is valid in case if quantum physics applies to the universe, and this question is 'another big if'. On the other hand, the quantum indeterminism is seems as indeterminism only from the point of view of classical mechanics, because quantum processes have no analogies in macro-world.

In general, there was a significant number of composers who used similar aleatoric methods and introduced degrees of indeterminacy in their compositional language; for example, German musicologist Thomas Schipperges examined (2011) the role of the aleatoric approach in unveiling Maderna's creative ideas in *Serenata per un satellite*, and Marina Pereverzeva (2013) provided a review of the mobile form in scores of different composers as well as an analysis of selected works of Roman Haubenstock-Ramati. Apart from that, Pierre Boulez's *Third Piano Sonata* is a vivid example of the implementation of aleatoric principles. Nathan M. Stephen Truelove deeply and mathematically analysed *Klavierstück XI* through the prism of a matrix system of serial polyphony and Stockhausen's method of translation of rhythmic structures into pitches (1984). As N.M.S. Truelove highlights that Stockhausen's matrix approach reminds similar systems that can be witnessed in the application of matrix algebra, and writes that 'Stockhausen employed three different types of matrices in constructing the six columns of the Final Rhythm Matrix: rhythm matrices, number matrices, and Roman numeral with exponent matrices' (p. 9). Truelove's analysis proves that Stockhausen not only possessed a formidable mathematical apparatus, but also successfully and effectively implemented it in music. As Truelove states, Stockhausen's original (preparatory) manuscript of *Klavierstück XI* contained 26 pages of sketches and notes, and these sketches demonstrate the methods used by Stockhausen to organise 'the rhythmic structure of *Klavierstück XI* before composing the pitch content of the piece' (pp. 3–4).

As Heisenberg's daughter Barbara Blum told me (during our meeting on 4 December 2023), her father Werner Heisenberg probably heard of the new attempts of Karlheinz Stockhausen. According to B. Blum (2020, pp. 134–138), in the 1960s there was an idea of creating a new Max Planck Institute for Music Research. One of the initiators of such an idea was distinguished biochemist Manfred Eigen, winner of the 1967 Nobel Prize in Chemistry, who came from a family of musicians and once even considered a career of a pianist. A special committee was gathered to discuss the new perspectives, included prominent scientists such as Werner Heisenberg and great musicians such as composers Pierre Boulez and Wolfgang Fortner, harpsichordist Edith Picht-Axenfeld, pianist Rudolf Serkin, violinist Yehudi Menuhin and baritone Dietrich Fischer-Dieskau, as well as philosophers Theodor W. Adorno and Georg Picht. Composition experiments and the development of a symbolic notation for new music were mentioned as specific tasks. In addition, music interpretations and investigations into chamber music styles, research into acoustics and finally electronic sound production and development of electronic instruments were planned. However, despite a promising start, the project did not last long and was not successful as planned. Boulez later went to found another institute in France (IRCAM). Werner Heisenberg and Pierre Boulez (as members of the committee) met regularly in Munich in the 1960s. Nevertheless, the views of Heisenberg (the oldest member of the committee) and Boulez (the youngest member) differed in a number of aspects. The goal of Boulez was to find new sounds and new instruments for music. Heisenberg's idea was that new instruments are not needed, but composers should go deeper into the questions of harmony and expressive capabilities of music. Heisenberg found the universe to be harmonious, for at the end the world was harmonious, and scientists had to explain the harmony of the world.

In general, Heisenberg's uncertainty principle is one of the most fundamental phenomena and pillars of quantum mechanics. As R. Feynman wrote: 'The uncertainty principle "protects" quantum mechanics. Heisenberg recognized that if it were possible to measure the momentum and the position simultaneously with a greater accuracy, the quantum mechanics would collapse' (2011, p. 138).

The explanation of Heisenberg's uncertainty principle given above is actually an oversimplification. 'The way you usually talk about the uncertainty principle is to discuss measurements of position and momentum. But it goes deeper than that. It is an intrinsic limitation on what position and momentum mean. Ultimately, position and momentum are not numbers.

They are more complicated objects called operators [...] What the uncertainty principle captures is not a lack of knowledge, but a fundamental fuzziness of the subatomic world' (Gubser 2010, p. 23).

Heisenberg himself once expressed the idea that 'This uncertainty relation specifies the limits within which the particle picture can be applied. Any use of the words "position" and "velocity" with an accuracy exceeding that given by equation (I) is just as meaningless as the use of words whose sense is not defined' (1930, p. 15).

By equation (I) Heisenberg refers to (p. 14):

$$\Delta x \Delta p_x \geq h$$

The first rigorous form of the equation was formulated by Earle Hesse Kennard in his article *Zur Quantenmechanik einfacher Bewegungstypen* (1927) and by Hermann Weyl in *Gruppentheorie und Quantenmechanik* (1928) as:

$$\sigma_x \sigma_p \geq \hbar/2$$

where $\hbar = h/2\pi$, and σ_x is the standard deviation of the position while σ_p is the standard deviation of the momentum.

Sometimes this equation appears as:

$$\Delta p \Delta x \geq \hbar$$

Δp is used for uncertainty in momentum, Δx stands for uncertainty in position, and $\hbar$ is the reduced Planck's constant (as already shown, $\hbar = h/2\pi$).

A slightly different version of this famous equation is used now (and can be found in, for example, Gubser 2010, p. 22 and Rae 2018, p. 12):

$$\Delta p \Delta x \geq h/4\pi$$

About 'words whose sense is not defined', Heisenberg writes in a footnote: 'In this connection one should particularly remember that the human language permits the construction of sentences which do not involve any consequences and which therefore have no content at all—in spite of the fact that these sentences produce some kind of picture in our imagination' (1930, p. 15).

During my meeting with leading philosophers of physics Jeremy Butterfield and Harvey Brown (22 November 2023), Jeremy told a joke about position and momentum: 'The joke that Harvey and I heard from Jeff Bub a delightful philosopher of quantum theory is: "The trouble with the philosophy of quantum mechanics is that once you find your position, you lose your momentum".'

The uncertainty principle is one of the foremost demonstrations of the deep indeterminacy of quantum mechanics; of course, it is 'indeterminate' only from the point of view of classical mechanics.

There is a famous story that happened at the legendary Fifth Solvay Congress on Physics that was held in Brussels in October 1927 (Fig. 13.2). Albert Einstein, though playing an important role in the beginning of quantum theory (with his photoelectric effect), strongly objected to the fundamental indeterminacy, refused to accept quantum probabilities, and

Fig. 13.2 Fifth Solvay Congress on Physics in Brussels, October 1927. This photo is sometimes called 'the most intelligent photo in the history of humanity' • Third row (from left to right): Auguste Piccard, Émile Henriot, Paul Ehrenfest, Édouard Herzen, Théophile de Donder, Erwin Schrödinger, Jules-Émile Verschaffelt, Wolfgang Pauli, Werner Heisenberg, Ralph Howard Fowler, Léon Brillouin.
• Second row (from left to right): Peter Debye, Martin Knudsen, William Lawrence Bragg, Hendrik Anthony Kramers, Paul Dirac, Arthur Compton, Louis de Broglie, Max Born, Niels Bohr.
• Front row (from left to right): Irving Langmuir, Max Planck, Marie Skłodowska Curie, Hendrik Lorentz, Albert Einstein, Paul Langevin, Charles Eugène Guye, Charles Thomson Rees Wilson, Owen Willans Richardson. Benjamin Couprie, Institut International de Physique Solvay, Brussels, Belgium
(*Source* Wikimedia Commons. Public domain)

said: God does not play dice. On this Niels Bohr replied: Albert Einstein, do not tell God what to do (Rae 2018, p. 48; Heisenberg 1983, pp. 116–117).

Actually, Werner Heisenberg did not like the term 'uncertainty'. He much more preferred the original German word *Unschärferelation*. Unschärferelation can be word-by-word translated as 'unsharp relations'. So for Heisenberg there was no uncertainty. As Barbara Blum said (during our meeting on 4 December 2023), the Copenhagen Interpretation of quantum mechanics was actually a 'political compromise' between Heisenberg and Bohr. This especially applies to the concepts of complementarity (one of Bohr's key ideas) and uncertainty relations. Once again, it is crucial to emphasize that Heisenberg himself did not like the term 'uncertainty', as for him the quantum reality was not 'uncertain', it was clear.

References

Adorno, T. 2018. *Aesthetics*, ed. E.Ortland, trans. W.Hoban. Cambridge: Polity Press.

Berry, M.V. 2023, June. Chaos, Quantum, Number. Submitted to *IoP Conference Series*.

Blum, B. 2020. Musik und Philosophie—Quellen der Kreativität bei Werner Heisenberg. In *Quanten 8*, ed. Kleinknecht, K. (ed.). Stuttgart: S. Hirzel Verlag.

Cage, J. 1960. *Solo for Sliding Trombone*. Edition Peters.

Cage, J. 1981[1992]. *Freeman Etudes*. Edition Peters.

Davies, P. 1989. Introduction. In *Physics and Philosophy*, W. Heisenberg, vii–xvii. Penguin Classics.

Davies, P. 2011. Introduction. In *Six Easy Pieces: Essentials of Physics Explained by Its Most Brilliant Teacher*, R. Feynman, R.B. Leighton, M. Sands, ix–xviii. New York: Basic Books.

Feynman, R., R.B. Leighton, and M. Sands. 2011. *Six Easy Pieces: Essentials of Physics Explained by Its Most Brilliant Teacher*. New York: Basic Books.

Gubser, S.S. 2010. *The Little Book of String Theory*. Princeton, NJ: Princeton University Press.

Hawking, S. 2022. *How Did It All begin? Brief Answers, Big Questions*. London: John Murray Publishers.

Heisenberg, W. 1930. *The Physical Principles of the Quantum Theory*, trans. C. Eckart and F.C. Hoyt. New York: Dover Publications, Inc.

Heisenberg, W. 1983. *Encounters with Einstein: And Other Essays on People, Places, and Particles*. Princeton, NJ: Princeton University Press.

Heisenberg, W. 2000. *Physics and Philosophy*. Penguin Classics.

Kasparov, Y. 2020. *Trombon. Evoliutsiya v XX veke i novye priemy igry* (Trombone. Evolution in XX Century and New Performing Methods). Moscow: Moscow State Tchaikovsky Conservatory.

Kennard, E.H. 1927. Zur Quantenmechanik einfacher Bewegungstypen. *Zeitschrift Für Physik* 44: 326–352.

Kleinknecht, K., ed. 2020. *Quanten 8*. Stuttgart: S. Hirzel Verlag.

Laplace, P.-S. 1902. *A Philosophical Essay on Probabilities*, translated from the 6th French ed. by F.W.Truscott and F.L.Emory. New York: John Wiley & Sons, London: Chapman & Hall.

Pereverzeva, M. 2013. Musical Mobile as a Genre Genotype of New Music. *Lietuvos Muzikologija* 14: 119–134.

Rae, A. 2018. *Quantum Physics: Illusion or Reality?*, 2nd ed. Cambridge: Cambridge University Press.

Schipperges, T. 2011. Bruno Madernas „Serenata per un satellite" (1969). Eine musikalisch inszenierte Huldigung. In *Inszenierung durch Musik. der Komponist als Regisseur* (Liber amicorum für Silke Leopold), hg. von Dorothea Redepenning und Joachim Steinheuer, Kassel.

Stockhausen, K. 1956. *Klavierstück XI*. Vienna: Universal Edition.

Stockhausen, K. 1969. *Momente for solo soprano, 4 choir groups and 13 instrumentalists*. Kuerten: Stockhausen-Verlag.

Stockhausen, K. 1989. *Stockhausen on Music: Lectures and Interviews Compiled by Robin Maconie*. London: Marion Boyars.

Truelove, N.M.S. 1984. *Karlheinz Stockhausen's Klavierstück XI: An Analysis of Its Composition Via a Matrix System of Serial Polyphony and the Translation of Rhythm Into Pitch*. DMA Dissertation. Norman: University of Oklahoma.

Vedral, V. 2018a. *Decoding Reality: The Universe as Quantum Information*. Oxford: Oxford University Press.

Vedral, V. 2018b. *From Micro to Macro: Adventures of a Wandering Physicist*. World Scientific Publishing Company.

Weyl, H. 1928. *Gruppentheorie und Quantenmechanik*. Leipzig: Hirzel.

14

Time Dilation, Synchronicity and Ligeti's Etudes

Very interesting analogies can be found in Steven S. Gubser's *The Little Book of String Theory* (2010), where the author compares certain phenomena of quantum mechanics with Frédéric François Chopin's *Fantaisie-Impromptu* C♯ minor, Op. posth. 66, WN 46 (Fig. 14.1) composed in 1834 and first published posthumously in 1855.

As Gubser wrote, 'The combination gives the composition an ethereal, liquid sound. It's a beautiful piece of music. And it makes me think about quantum mechanics. […] In quantum mechanics, every motion is possible, but there are some that are preferred. These preferred motions are called quantum states. They have definite frequencies. A frequency is the number of times per second that something cycles or repeats. In the Fantasie-Impromptu, the patterns of the right hand have a faster frequency, and the patterns of the left hand have a slower frequency, in the ratio four to three. […] Simple quantum systems, like the hydrogen atom, have frequencies that stand in simple ratios with one another. For example, the phase of one quantum state might cycle nine times while another cycles four times. That's a lot like the four-against-three cross rhythm of the Fantasie-Impromptu. But the frequencies in quantum mechanics are usually a lot faster' (Gubser 2010, pp. 19–20).

The very principles of these analogies are useful, however, the comparison with Chopin's *Fantasie-Impromptu* is exceedingly simple and basic. The main factor of the utter simplicity of this comparison is that from contemporary composers' point of view Chopin's music is utterly 'Newtonian' in motion and mechanics and 'Euclidean' in terms of space.

R.-B. Abdyssagin, *Quantum Mechanics and Avant-Garde Music*, https://doi.org/10.1007/978-3-031-63161-0_14

Fig. 14.1 Beginning of the score of Frederic Chopin, *Fantaisie-Impromptu* in C♯ minor, Op. posth. 66, WN 46 (Edited by Herrmann Scholtz. Samtliche Pianoforte-Werke, Band II (pp. 331–36). Publisher: Leipzig: C.F. Peters. Public Domain)

Using the same principle as S. Gubser used, far greater poetic similarity can be found providing comparison of quantum mechanics and contemporary music, for instance, taking complex rhythmic structures in a cycle of *18 études* for piano by Gyorgy Ligeti (1985–2001). In *Étude 1: Désordre* the rhythmic synchronisation between right and left hands is developed to a totally new level: the order becomes disorder (chaos) and disorder (chaos) symbolizes order. In fact, two hands seem to exist in two totally different realities, dimensions, initially being born from one, they emerge to represent diverse layers of synchronicity. In fact, synchronicity is almost equalised with asynchronous movement (a-synchronicity). Not because they become one or same thing, not at all. But each new synchronicity produces new dimensions of asynchronous movements, and this expansion grows steadily like a rhythmic/metric entropy of the work, and finally reconciles and explodes in the last bar. In this étude every passage or musical figure indeed can be figuratively compared with quantum systems. In Chopin's *Fantasie-Impromptu* the 'faster' movement is 'faster' only in imagination, while Ligeti's masterful method elevates the relational force of synchronicity/a-synchronicity to a new level. This proves to be one of the most profound differences not only between two masterpieces and their creators (Chopin and Ligeti), but also between classical/romantic tradition and avant-garde music. This difference in music is no less profound than the difference in physics between classical (Newtonian) mechanics and quantum mechanics.

In *Étude 2: Cordes à vide*, as well as *Étude 5: Arc-en-ciel* despite the illusion of formal synchronous movement, the poly-rhythmic aspects are greatly disguised within strict and stable bars. In fact, the richness and independence of patterns and passages may be compared to the effect of time dilation, which is described by S. Gubser as 'You can move any direction you want in space, but you can't move backward in time. In fact, you can't really "move" in time at all. [...] But it's actually not that simple. If you run in a circle really fast while a friend stands still, time as you experience it will go by less quickly. If you and your friend both wear stopwatches, yours will show less time elapsed than your friend's. This effect, called time dilation, is imperceptibly small unless the speed with which you run is comparable to the speed of light' (2010, p. 12).

In Ligeti's *Étude 1: Désordre*, *Étude 2: Cordes à vide* and *Étude 5: Arc-en-ciel* each hand (or, better say, dimension) moves with its own speed, and poetically causes the effect of dilation between them. I would introduce the term of multi-synchronicities, because the formal synchronous movement is retained as time is a physical dimension, but 'particular' or 'special' synchronicities are not 'synchronous', and this peculiar correlation of 'synchronicities' may be called multi-synchronicities. Putting simply, in *Étude 2: Cordes à vide* and *Étude 5: Arc-en-ciel* each musical voice (in polyphonic and textural senses) is aligned with others formally within the metric system of composition, which causes a formal synchronicity. They are not really a-synchronous per se. But when each voice (while being formally synchronous) has its own musical imaginary speed and rhythmic space, this leads to the emergence of multi-synchronicity. This example displays the existence of multiple synchronicities within the formal structure of rhythmic alignment of bars. The notion of multi-synchronicities can be successfully applied to many examples of contemporary music pieces, and Ligeti's *études* are simply among the most obvious ones. I would say that co-existence and structural implementation of several multi-synchronicities can lead to emergence of a concept of super-synchronicity (thus, super-synchronicity is a result of superposition of several multi-synchronicities).

Using Gubser's method of analogies, it is even possible to provide a distant correlation between Ligeti's compositional technique in these études and the famous 'twin paradox'. The 'twin paradox' is brilliantly explained by R. Feynman: 'We consider a famous so-called paradox of Peter and Paul, who are supposed to be twins, born at the same time. When they are old enough to drive a spaceship, Paul flies away at very high speed. Because Peter, who is left on the ground, sees Paul going so fast, all of Paul's clocks appear to go slower, his heartbeats go slower, his thoughts go slower, everything goes slower, from Peter's point of view. Of course, Paul notices nothing unusual, but if he travels around and about for a while and then comes back, he will be younger than Peter, the man on the ground! That is actually right; it is one of the consequences of the theory of relativity which has been clearly demonstrated' (2011, pp. 77–78).

Coming to this analogy, Ligeti and many other notable contemporary composers used methods of juxtaposing different rhythmic velocities and paces in otherwise synchronous music works. Ligeti's *études* are masterpieces demonstrating how right and left hand can move with different speeds, and as a result, each hand will have its own 'time'—both musically and philosophically. Implementing Gubser's metaphoric method of comparisons, in Ligeti's *études* the hands of a piano performer act almost like twins in the 'twin paradox': each moves at different speed and finally demonstrates different 'age'. 'Age' only means that different hands may proceed through different stages of musical form at the same time; they are formally synchronous because they sound simultaneously, but their 'location' within dramaturgy can be deemed as different. This can be named an effect of multi-synchronicity or even super-synchronicity.

This phenomenon is present not only in Ligeti, but also in the scores of many other leading composers. Rhythmic superpositions are widely featured in the works of composers of the direction of new complexity. James Erber's piano piece *Qfwfq* (2003) is a clear example of this (Fig. 14.2).

Actually, this analogy is also very fragile, because in the 'twin paradox' one of the twins should move with the speed of light, and this (as well as another important aspects) is impossible in music even on a metaphoric level. Thus, the mentioned examples with Ligeti's études serve only to demonstrate how

Fig. 14.2 Excerpt from the score of James Erber *Qfwfq* (© James Erber)

Gubser's method of drawing analogies can function fruitfully in the context of contemporary music.

As additional effective examples that can suite Gubser's method of analogies, it is possible to mention works of Olivier Messiaen, Iannis Xenakis, Bruno Moderna, Pierre Boulez, Karlheinz Stockhausen, Luciano Berio, Luigi Nono, Witold Lutosławski, Roman Haubenstock-Ramati, Wolfgang Rihm, Brian Ferneyhough, Gerard Grisey, Tristan Murail, Hugues Dufourt, Philippe Hurel, Franco Donatoni, Salvatore Sciarrino, George Crumb, Gyorgy Kurtag, Ivan Fedele, Alessandro Solbiati, Georges Aperghis, Georg Friedrich Haas, Beat Furrer, Chaya Czernowin, Mark Andre, Rebecca Saunders, Enno Poppe, Alberto Posadas, Carola Bauckholt, Simon Steen-Andersen and many many others.

In fact, almost every respectable avant-garde composer pays essential attention to rhythm, and poly-rhythmic structures (incomparably far more complex than those present in classical and romantic traditions) appear in the compositions of many generations of contemporary composers.

References

Chopin, F. 1894. *Fantaisie-Impromptu* Cis-moll, Op. posth. 66, WN 46, ed. Herrmann Scholtz. Leipzig: C.F.Peters.

Erber, J. 2003. *Qfwfq* for solo piano. Composers Edition.

Feynman, R., R.B. Leighton, and M. Sands. 2011. *Six Not-So-Easy Pieces: Einstein's Relativity, Symmetry, and Space-Time*. New York: Basic Books.

Gubser, S.S. 2010. *The Little Book of String Theory*. Princeton, NJ: Princeton University Press.

Ligeti, G. 1985–2001. *Études pour piano*, Books I, II and III. Mainz: Schott Musik International GmbH & Co.

15

Pauli Exclusion Principle and Schoenberg's Dodecaphony

According to the exclusion principle formulated by Wolfgang Pauli, in one quantum state there can be no more than one fermion. Distantly analogous to this, in music, the dodecaphonic principle of series (tone rows) forbids the existence of the same note (i.e. the same note/pitch repeated in different octaves and/or under other circumstances counts as identical; a pitch class) within the same series (tone row). The octave (equal temperament within the framework of classical music) has 12 tones. Within the framework of the twelve-tone technique 1 note cannot be repeated until the remaining 11 are performed. Thus, the in-principle existence of the same/identical note within the same series (tone row) is excluded.

So, if to metaphorically consider a tone row in dodecaphony as a quantum state in quantum mechanics, and if to consider a single note in music as a fermion in the micro-world, then we have quite a transparent imaginary correlation. It is breathtaking indeed to realise that the exclusion principle was formulated almost at the same time as the very first dodecaphonic music was composed: 1925 for Pauli's exclusion principle and 1921/23 for Schoenberg's *Suite for Piano* Op. 25!

That being said, we should not get overexcited and try to implement this analogy wider than it is present now. One of the first fundamental limitations in the explained correlation lies in the fact that while Schoenberg's dodecaphonic principle of impossibility of existence of more than one identical note (pitch class) within the same tone row applies to all tones/notes without exceptions, Pauli's exclusion principle is valid only for fermions (particles with half-integer spins), whereas bosons (particles with integer spins) can easily occupy the same quantum state in any quantity.

R.-B. Abdyssagin, *Quantum Mechanics and Avant-Garde Music*,
https://doi.org/10.1007/978-3-031-63161-0_15

Additionally, the twelve-tone technique evolved into the serialism of Webern and, subsequently, to the total (or integral) serialism of Pierre Boulez (1925–2016), where the principle of rows/series was extended to cover not only tones/pitches, but also rhythmic durations, dynamics, timbral characteristics and numerous other parameters of music composition (Boulez, 1950).

Furthermore, all subatomic particles must be either bosons or fermions, while the very possibility of division of tones/notes into categories within the context of the twelve-tone technique seems quite meaningless. Of course, the notion of consonances and dissonances as specific relations of notes forming intervals was known, although their relative content has been evolving and constantly changing throughout the history of music. But one of the foremost achievements of Schoenberg's dodecaphony was the complete obliteration and negation of the interrelation of consonances and dissonances, for the dissonances were 'liberated', and there were no longer any types of intervals.

P. Hindemith was eminent in the creation of his own harmonic system with levels of tension between intervals, which was a step forward after dualism of consonance and dissonance. Hindemith even had a notion of 'the nature of the atomic structure of music' (1937/1941, cited in Pareyon 2011, p. 130).

But even this breakthrough has little-to-no relation with bosons and fermions. Thus, the lack of any types of intervallic relation within the context of Schoenberg's twelve-tone technique remains a vivid example of the borders of its poetic connection with Pauli's exclusion principle. Even more, the fact that Schoenberg's dodecaphony consists only of 'fermions' and lacks any 'bosons' means that the twelve-tone technique has a fundamental limitation. In the sense that Schoenberg's dodecaphony does not encompass the entire musical universe; quite the contrary, the twelve-tone technique is essentially one of the steps (limited in its nature) within the infinite evolution and progress of the art of music.

Historically, there have been numerous attempts to imagine several 'gravitational points' or 'centres of force' within one tonal or modal system and event intentions to invent new systems that would function independently and out of the field of influence of traditional tonality—specific works by Josef Matthias Hauer (1883–1959), Fritz Heinrich Klein (1892–1977), Alexander Scriabin (1872–1915), Nikolai Roslavetz (1881–1944) (Fig. 15.1), Nikolai Obukhov (1892–1954), Arthur Lourié (1892–1966), research of Georgy Lvovich Catoire (1861–1926) and many others can serve as examples (to a certain extent).

Fig. 15.1 Excerpt from the score of N. Roslavets *Piano Sonata № 1* composed in 1914, published in 1990 in Moscow: Muzyka

J.M. Hauer (1920) and F.H. Klein (1921) are believed to unfold the opportunities present in emerging twelve-tone frameworks and 'ex-tonal' compositional concepts even slightly earlier than Schönberg. The first steps in exploring the new systematic possibilities of not only the tonality, but even notation took place, for example is A. Lourié's *Formes en l'air* composed in 1915 (Fig. 15.2).

N. Obukhov's twelve-tone chromatic notation totally eliminates the very essence and any need in accidentals, traditional keys, tonalities and previous understanding of harmony, because all notes increased by half-tone are indicated by crosses in his notation. Stating simply on the example of piano keyboard, he uses ordinary noteheads for seven natural ♮ diatonic keys—C, D, E, F, G, A, H—and crosses ✖ for half-step interval keys—C♯/D♭, D♯/E♭, F♯/G♭, A♯/B♭ (Obukhov 1943, 1948, 1953). Interestingly enough,

Fig. 15.2 Excerpt from A. Lourié *Formes en l'air*. Moscow: Muzgiz

this original method of notation does not deprive its author's music of clear signs of poly-tonal/modal thinking with evident 'centres' of attraction. Worth mentioning that the evolution of new music also leads to advancements in the terminology of music theory—one of the bright examples is the appearance of the concept of 'contonation' (not to be confused with intonation or connotation).

Concluding, it is important to state the most fundamental aspect concerning the relations between the Pauli exclusion principle and Schoenberg's dodecaphony: the Pauli exclusion principle exists in the nature, and Wolfgang Pauli was the first to discover and explain it. Speaking metaphysically, this principle was created by God and cannot be changed or altered. In contrast, the guiding principles of dodecaphony are not part of natural laws, they are created by human being—invented by Arnold Schoenberg.

Dodecaphonic implementation may be amended, altered and changed on a wish of every composer who desires to uses it. Although the very possibility of twelve-tone (not in conventional tonality) music always existed in nature, dodecaphony as a system was an invention. So, as was already stated, Schoenberg was both a discoverer and inventor at the same time: he was one of the first to see (discover) the existing (in nature) potential of creating music beyond the scope of the classical tonal system, but specific methods and principles of dodecaphony were his inventions.

The Pauli exclusion principle is a scientific law 'created by God' and all fermions obey it. On the other hand, Schoenberg's principles are relative and valid only within his own framework, and he himself many times 'disobeyed' these principles (even after manifesting them). If God created the universe and all existing natural laws and every object in his universe has to obey and follow them (as there is no other way, except for the existence of another law that permits it), in musical composition all the 'laws' are created by a composer, and only he decides what range and limitations these 'laws' will have.

Here we identify a deep difference between a scientist and a composer. 'At best, a scientist can be a discoverer—the one who first sees and explains already existing phenomena in nature, and the objective laws of nature cannot be changed or corrected. A composer is a creator, if he does not write this music, there is a chance that this music will never be written' (Abdyssagin 2017a, p. 44).

On one occasion even Heisenberg himself said 'If I had never lived, someone else would probably have formulated the uncertainty principle. If Beethoven had never lived, no one would have written op. 111' (Gingerich 1975, cited in Crease and Pesic 2014).

Considering the discussion about creation, in 2016 I composed a symphonic picture *God's Dwelling* (Fig. 15.3) for full orchestra (2017b). The premieres of this orchestral work were held in Berliner Konzerthaus—Großer Saal, Rudolfinum Prague, Wiener Konzerthaus—Großer Saal, and Guildhall London, in August-November 2016. The epigraph of this composition is:

We think that God sees us from above.
But He knows us from inside.

Fig. 15.3 Beginning of the score of Rakhat-Bi Abdyssagin *God's dwelling* (2016) (© Rakhat-Bi Abdyssagin 2016. All Rights Reserved)

References

Abdyssagin, R.-B. 2017a. *Bozhestvenniy put'* (The Divine Path). Almaty: Kazak Universiteti.

Abdyssagin, R.-B. 2017b. *God's Dwelling* symphonic picture for full orchestra. Moscow: Kompozitor Publishing House.

Boulez, P. 1950. *Piano Sonata No. 2.* Paris: Heugel et Cie.

Crease, R.P., and P. Pesic. 2014. Physics and Music. *Physics in Perspective.* 16: 415–416.

Hauer, J.M. 1920. *Vom Wesen des Musikalischen.* Leipzig/Wien: Waldheim-Eberle.

Klein, F.H. 1921. *Die Maschine: Eine extonale Selbstsatire* für Klavier zu 4 Händen, Op.1. Vienna: Carl Haslinger.

Lourié, A. 1920. *Formes en l'air* pour piano. Moscow: Gosudarstvennoe muzykal'noe izdatel'stvo.

Obukhov, N. 1943. *Aimons-Nous les uns les Autres pour piano*. Paris: Éditions Durand.

Obukhov, N. 1948. *La paix pour les réconciliés* pour piano. Paris: Éditions Durand.

Obukhov, N. 1953. *Le temple est mesuré, l'esprit est incarné* pour piano. Paris: Éditions Durand.

Pareyon, G. 2011. *On Musical Self-Similarity*. Imatra: The International Semiotics Institute.

Roslavets, N. 1990. *Piano Sonata No 1*. Moscow: Muzyka.

Schoenberg, A. 1925. *Suite for Piano*, Op. 25. Vienna: Universal Edition.

Part III

Shadows of the Void

16

Shadows of the Void

The lack of 0, or absence of void is one of the essential characteristics of early western thinking. Ancient Greeks did not have 0 (Vedral 2018, p. 32), and it is also absent in Roman numerals. However, the Arabic numerals contain 0 as one of their fundamentals. 0 came to the Arabic numeric system from India and Persia, and this crucial difference between Arabic (Eastern) and Roman (Western) numerals also somehow displays the different civilisational aspects of the Eastern and Western worlds at that time. This phenomenon also explains why algebra and ancient calculus have initially been born and developed in the East, and only later became inseparable parts of the emerging Western thinking. The mentioning of only al-Khwarizmi is already a brilliant example, and it is believed that the word 'algorithm' is the transliteration of his name. This passage serves purely as a means of demonstrating how important was and remains the discovery and perception of 0 and Void.

In a number of my own musical works (starting from 2011) the silence in music is treated as 0 in mathematics and the void in physical space. In my book *Mathematics and Contemporary Music* (2013, p. 12) (Fig. 16.1), I discovered the following metaphorical correlation: Silence—0—Void. The tremendous impact and an overarching importance of void is obvious. As Stefan Zweig wrote, '…nothing on earth exerts such pressure on human soul as a void' (2017, p. 40).

The Void is not nothing, the Void is less than nothing but more than everything. Speaking philosophically, the Void, or 'what does not exist', is more important than 'what exists', because what does not exist allows the existence of what exists.

R.-B. Abdyssagin, *Quantum Mechanics and Avant-Garde Music*, https://doi.org/10.1007/978-3-031-63161-0_16

Fig. 16.1 Cover of my book *Mathematics and Contemporary Music* (2013), where I discovered the metaphorical correlation: Silence—0—Void

Comparing the role of silence in music to the role of the void in the universe, it is worth noting that Dark Energy and Dark Matter shape the universe on the largest scale. As I wrote in a programme note to my composition *Ombre del Vuoto* (Shadows of the Void) for 11 performers: Emptiness or the void is not just a physical idea but it is also a philosophical one. Anything that goes beyond human's touch or reach is considered to be some emptiness or a void. But is it a real void?! In reality there are many ambiguities and phenomena that cannot be understood or perceived, which we tend to call a 'void'. These 'voids' also do not go for good and are not left unnoticed. Certain traces, signs and symbols from these 'voids' are still there to influence the formation of the universe in a powerful manner as anything material does. Just like a physical person's activities cast shadows, the results of the 'voids' also cast shadows which intertwine and create new realms of 'space and time'. These realms are the *Shadows of the Void* (Abdyssagin 2019).

The world premiere of this composition was held in Rome (Italy), Accademia Nazionale di Santa Cecilia, Auditorium Parco della Musica, Teatro Studio, by Ensemble Nocevento on 21st June 2019.

Here I provide the introductory structural analysis (Abdyssagin 2022, pp. 101–103) of *Ombre del Vuoto* (2019):

11 performers, 14 instruments (one percussionist on several drums). The basis of the harmonic field lye in the tetrachord: F♯, G♯, H, C. This is one of the omnintervallic tetrachords—4-sound-chords that contains all the intervals (including their inversions). Global *campo armonico* (harmonic field) of this score is created by two symmetrical chords (which together in different situations can even be considered a scale or distant analogue of a mode), each of which includes 6 sounds.

The first chord: C, D♭, E, F♯, A, B♭. This chord consists of two tetrachords (containing all intervals as well)—C, D♭, E, F♯, as well as E, F♯, A, B♭… It is worth mentioning that E and F♯ are common for both tetrachords and form the core of the given chord.

The second chord: C, D, F, G♭, A, H. This is the mirror image of the first chord, and it also consists of two tetrachords: C, D, F, G♭, and F, G♭, A, H.

Both of these chords are palindromes—if you read them from the top to the bottom or from the bottom to the top—they will have the same intervals! These two symmetrical chords (each containing 6 tones), superimposed on each other (taking into account the overlapping sounds), created a new *super*chord of 9 tones. This *super*chord became the starting point in determining the pitch and frequency interactions inside the piece. 9 tones correlate to the harmonic environments of this score. The 9 given tones are as follows: C, D♭, D, E, F, F♯, A, B♭, H.

There are 9 phases in the development of this piece. Their intersections and additions are determined by sonoristic and metro-rhythmic aspects. For example, a flute may have the 8th developmental phase, while a clarinet in this place may have the 9th developmental phase, and another instrument may have the 7th developmental phase. Out of the 12 tones (the standard chromatic scale within an octave, excluding microtones or quartertones), 9 are used as the fundamental chord, and each of the 9 developmental phases is dedicated to the spectrum of one tone. And the 3 remaining tones are considered as 'out-of-phase' and are used (mainly the tones of their spectrum) freely during cadence-like movements. In the very beginning, the first phase is built around the F sharp spectrum since this tone is present in both chords and is located in the center of the general 9-tone scale of the *super*chord (from C).

Nevertheless, this 9-tone *super*chord is not fully present anywhere in the score! It is absent, being quasi-void. But this 9-tone *super*chord pre-defined and determined almost everything else in the structure of this composition. It is the foundation of the formal language of this sinfonietta. This 9-tone *super*chord 'does not exist' in the score but rules everything that 'exists' in the score. Which is very much like the Void. As already stated above, the Void, or 'what does not exist', is more important than 'what exists', because what does not exist allows the existence of what exists. Thus this piece becomes one of the possible implementations of this principle. However, there are of course many other possible ways of realising this idea.

Starting from the first page (Fig. 16.2) of the score of *Ombre del Vuoto*, the example of notating multiphonics is seen, which indicates not only the fingering but also the dynamics, position of the lips and pressure, which gives the performer a complete and comprehensive idea of how it should be played. In *Ombre del Vuoto*, the fingerings of the multiphonics on the flute derive from the treatise by Pierre-Yves Artaud (1995), for the oboe from the treatise by Peter Veale (1994), and for the bassoon from the treatise by Pascal Gallois (2009).

In this work, almost all the expressive capabilities of the instruments known at the moment are used systematically, in accordance with the original plan. Apart from the overall 'strategy', locally many individual 'tactical' concepts were created in order to fit and meet the basic idea. The extended techniques of the instruments used are inextricably intertwined with the 'layer of frequencies' for which a full range of the latest composing techniques are applied, even those that come from electronic music, for example, a *harmonizer* and *filtraggio* (filter) (Fig. 16.3).

Here are some clarifications on what the *campo armonico* (harmonic field) is; I would like to demonstrate one of the possible iterations of this concept based on the example of Maestro Ivan Fedele. We can consider two separate combinations of 4 tones within the interval of fifth.

The first combination (tetrachord): C, C♯, D, G.
The second combination (tetrachord): C, F, F♯, G.

The most important and key parameter of these two given tetrachords is that neither of them consists of any interval of third and/or sixth in the interaction between the sounds—these tetrachords can be varied according to well-established polyphonic techniques, while retaining their basic characteristics.

Fig. 16.2 Beginning of the score of Rakhat-Bi Abdyssagin *Ombre del Vuoto* (2019) (© Rakhat-Bi Abdyssagin (2019). All Rights Reserved)

Fig. 16.3 Excerpt from the score of Rakhat-Bi Abdyssagin *Ombre del Vuoto* (2019)
(© Rakhat-Bi Abdyssagin (2019). All Rights Reserved)

However, the most important tetrachords among the composer's methods
and 'instruments'—being very significant ones—are those that contain all
the intervals if we take into account inversions of the intervals as well (like
the mirrored intervals), as shown above. It is obvious and well known that a
minor second is the 'mirror' (inverted version) of a major seventh. A major
second is the inversion of a minor seventh. It is quite the same with thirds
and other intervals.

We can deduce 4 similar fundamental combinations (omnintervallic tetrachords):

First: C, Db, E, F♯.
Second: C, D, F, Gb.
Third: C, Db, Eb, G.
Fourth: C, E, F♯, G.

The first and the second tetrachords are within the interval of a tritone,
while the third and the fourth are within the fifth. Moreover, the second
and the fourth are a 'mirror-kind' of reflections of the first and the third
respectively. And these are the 4 original tetrachords that contain all intervals
(including inversions). Labelling them with numbers is purely arbitrary, and

does not imply any sequence of importance or anything similar; it was done just to simplify the explanation.

In *Ombre del Vuoto* I implemented complex methods and approaches to frequency organisation—being based on imagination—and showed their possible correlations with sonoristics and contemporary techniques.

Fundamental aspects of the theoretical framework of the reviewed omnintervallic tetrachords and practical tools for their implementation in music compositions have been developed and demonstrated by Maestro Ivan Fedele during his lectures at the Accademia Nazionale di Santa Cecilia in Rome, Italy (2017), where I studied *Corso di Perfezionamento in Composizione* (equivalent of an artistic doctorate/PhD in music composition) from 2017 until 2019.

References

Abdyssagin, R.-B. 2013. *Mathematics and Contemporary Music*. Almaty: Kazak Universiteti.

Abdyssagin, R.-B. 2019. *Ombre del Vuoto* per 11 esecutori ('Shadows of the Void' for 11 performers).

Abdyssagin, R.-B. 2022. *Noveishie kompozitorskie i ispolnitel'skie tehniki* (The Newest Composing and Performing Techniques). Astana: KazNUA.

Artaud, P.-Y. 1995. *Flûtes au présent/Present Day Flutes*. Paris: Billaudot.

Gallois, P. 2009. *The Techniques of Bassoon Playing / Die Spieltechnik des Fagotts/La technique du basson*. Kassel: Bärenreiter-Verlag.

Veale, P., C.S. Mahnkopf, W. Motz, and T. Hummel. 1994. *Die Spieltechnik der Oboe/The Techniques of Oboe Playing/La technique du hautbois*. Kassel: Bärenreiter-Verlag.

Vedral, V. 2018. *Decoding Reality: The Universe as Quantum Information*. Oxford: Oxford University Press.

Zweig, S. 2017. *Chess*, trans. A. Bell. Penguin Classics.

17

Boolean Algebra, Bits and Qubits in Music

The mathematician George Boole (Fig. 17.1) developed unique algebraic approach, stating that all human logic can be reduced to the manipulations of 0 and 1 (1854). This became the foundations of what would later be called as the 'information age', so Boole's role as one of the founding fathers of computer science is widely acknowledged. He demonstrated that all possible algebraic operations and manipulations can actually be done by using only two numbers: 0 and 1. A digit that is either 0 or 1 is named a 'binary digit', this is why the word 'bit' (shortened version of 'binary digit') appeared.

Symbolically, the notation of classical music (let's say from early baroque to the beginning of the twentieth century) can also be expressed by the means of 0 and 1, when 1 is sound and 0 is silence. These phenomena of silence and sound have been forming the musical thought over centuries, and essentially remain the kernel of classical music. Therefore, we can say that 'bits' also exist in composition and are present in classical forms of notation. As the information is physical, the most important carriers of information in music are precisely the silence and a sound.

In modern days, quantum reality 'upgraded' the classical notion of a bit to 'qubits' (quantum binary digit). 'A qubit is a quantum system that, unlike a bit, can exist in any combination of the two states, zero and one' (Vedral 2018, p. 131). In contemporary music, owing to the large number of extended methods of playing instruments, there are sonoristic effects that create sounds that poetically exist in peculiar combination of silence (0) and sound (1). Among multiphonics, sons fendus, harmonics, bisbigliando, slap tongue, tongue ram, key clicks, jet whistle, whistle tones and others, we specifically note the techniques of air noise, half tone/half noise, aeolian

R.-B. Abdyssagin, *Quantum Mechanics and Avant-Garde Music*, https://doi.org/10.1007/978-3-031-63161-0_17

Fig. 17.1 G. Boole (*Source* Wikimedia Commons)

sounds, soffio and the likes that are capable of representing utter expressivity, mystical and magical sonoristic beauty as well as charming states of tones transitioning between silence and sound.

As a simple demonstration of how music pieces can be structured using these new techniques in 'superposition' of 0 and 1, I will provide structural analysis (2022a, pp. 72–76) of my composition *Light through fog* for flute solo (2012, 2022b). It was composed between 5th and 8th November 2012, when I was 13 years old, and was essentially a study of how to write a solo flute piece implementing only and exclusively the extended performing methods.

First, the system of notating musical time was determined—in the given case using seconds. There are different ways to capture time in music. Regarding 'seconds' systems, the most common is the end-to-end method, which is used in *Serenata di stelle invisibili* (2019) and has already been analysed in Chapter 11. In the case of *Light through fog*, instead of a 'general timeline', an approach was used when the local designations of the durations

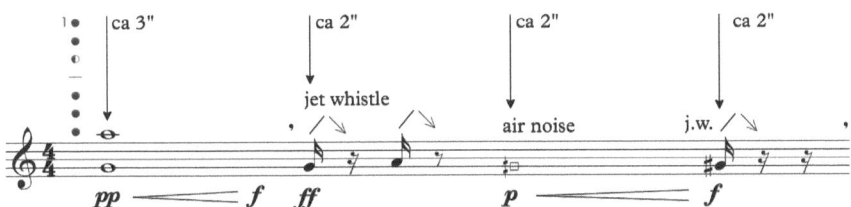

Fig. 17.2 Beginning of Rakhat-Bi Abdyssagin *Light through fog* (2012), published by Verlag Neue Musik Berlin

of each individual event (element) were given (Fig. 17.2). It is important to understand that such a system of capturing durations may be suitable in the case of works for solo instruments or for a small chamber ensemble, where each instrumentalist can play from the score. However, in the case of large ensemble work, and even more so with orchestral scores, where everyone has their own part, it is much more practical to use the traditional method of notating rhythm/durations/musical time, as this will speed up and significantly increase the efficiency of rehearsals.

A multiphonic sounds at the very beginning of *Light through fog*, followed by the jet whistle technique (an energetic 'whistle-impulse' from the indicated notes). Then comes air noise, which turns back into jet whistle. Let us pay attention to the formation of pitch. The first line uses the following notes: G, G quarter-sharp, G♯, A—a cluster between the sounds G and A, which were defined by the pitches of the first multiphonic.

Each line of the work has not only an internal degree of completeness in terms of dramaturgy, but also a certain amount of symmetry in terms of elements. For example, the sequence of durations of the first line in seconds: 3', 2', 2', 3' (10' together). A 'mirror' relationship arises, which also affects the general perception.

In a certain sense, the art of composition is the ability to make a lot out of little. The entire first phase of the development of the work (the first 1.5 pages of the score) is a series of combinations of techniques exhibited in the very first line (multiphonics, jet whistle, air noise).

In the second line (Fig. 17.3) each element appears twice in a row: two multiphonics (more precisely, one multiphonic with different fingerings), two jet whistles and two air noise.

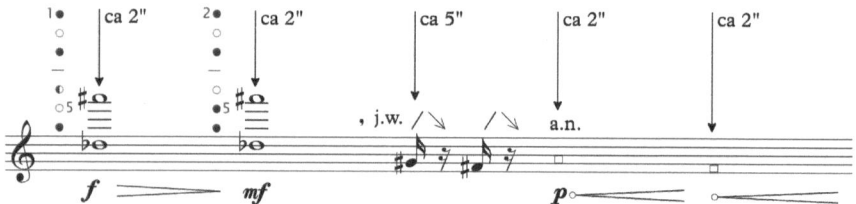

Fig. 17.3 The second line

The third line (Fig. 17.4) presents concentrated work with the jet whistle and air noise techniques (with the active use of microtones).

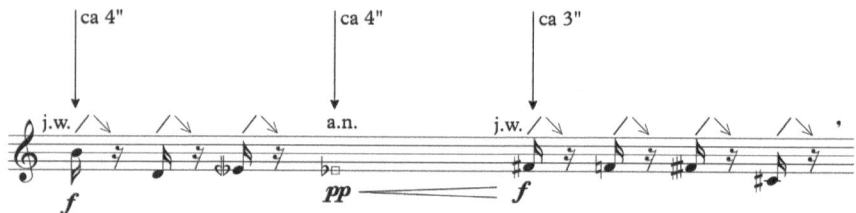

Fig. 17.4 The third line

In the middle of the second page, according to the dramaturgical plan, the previous elements in the process of interactions give rise to a new quality, which is embodied in a cascade of multiphonics (sustained), performed on a tremolo (and with a trill in the finale, Fig. 17.5).

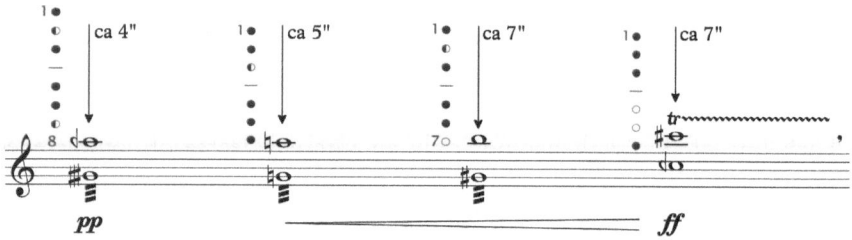

Fig. 17.5 Culmination of the first phase

The second phase (Fig. 17.6) is entirely devoted to all possible developments of the air noise technique. If in the first phase air noise was an auxiliary technique, while in the second phase it becomes the main one. From the point of view of the development of form, a dialectic occurs between the techniques of air noise and whistle tones (airy and soaring harmonics that sound very quiet). Gradually, frullato, hrullato (a special type of frullato, reminiscent

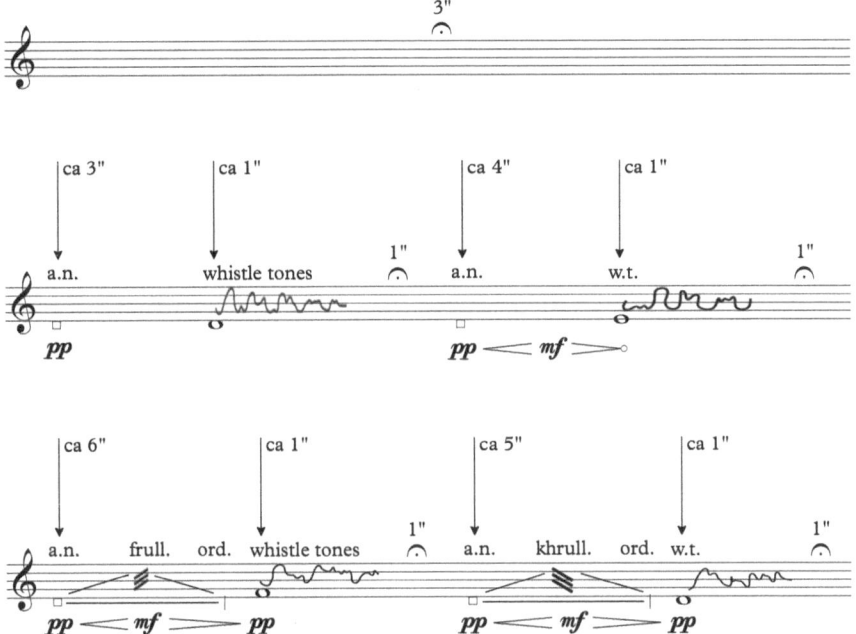

Fig. 17.6 The second phase begins after pause

of growth technique), smorzato are added to air noise. These elements increasingly accumulate, which at the end of the second phase leads to a maximum climax. Actually, this is where the concept of a single line from the quietest sound to the loudest climax comes into play. This dramaturgical line is easy to trace, since it is the core of this section of the form.

Pauses between elements not only allow the performer to take a breath, but also symbolise a mystical escape into other spaces and dimensions. This gives rise to a polyphony of layers and textures, the musical fabric itself begins to breathe and come to life.

In the third (final) phase (Fig. 17.7) of the work maximum attention is given to percussive effects—slap tongue, tongue ram, key noise, etc. Thus

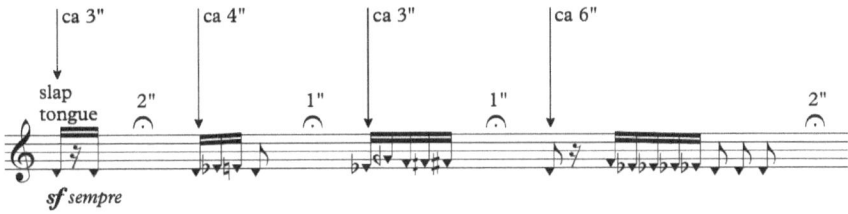

Fig. 17.7 The start of the third phase

phase starts with slap tongue on note D, which creates continuity between sections of the form, since the system-forming sound of the second phase was also note D. Considering that there are no bars in the work, accidentals, if necessary, are placed before each note and repeated.

Gradually, the end-to-end development of percussive techniques turns into multiphonics, which alternate with tongue ram. On the last page the sounds dissolve into silence. The work ends with a long and 'gentle' stream of whistle tones.

In the opus in question there is not a single note or sound extracted by the 'classical/traditional' method. The architectonics of the work is entirely based on sonorous sounds and advanced flute playing techniques.

Each technique in *Light through fog* is a reflection of a clear musical thought, a carrier of an artistic image.

Now let us consider one of the possible semantic interpretations of what happens in *Light through fog* (Abdyssagin 2013, pp. 18–19). The basis of dramaturgy lies in the contrast between combinations of multiphonics and jets and air noise. How can multiphonics be interpreted? In this particular case they are double notes, a kind of deformation of the ordinary flute sound. Improvising, we can review one of the possible versions of imaginary interpretation. Like the sun, many times reflected in the shards of a broken mirror, a regular flute sound is distorted in the split space and is reflected in its parts. Whether it is really the space distorts normal sound of the flute, or it is all right with space, but the flute is in some extreme conditions and its response is similar if not to a cry, but, in any case, to the distorted voice—one way or another, at the very beginning there is an intrigue, necessary for any composition.

This is followed by strong jets—with great energetic message—which sound like a lash, and a low whistle of air as a symbol of reaction to a pain, or as a threat, like a hissing snake. Thus, the play begins with a deformed, split flute sound, strong jets as slash or a scourge, and concealed hissing. This range of images served as a classical dualism in an understandable and familiar form of 'question–answer'. These images immediately begin to develop. Thus, regardless of the emotional interpretation of techniques, we deal with bright, imaginative, contrasting methods, and they are quite similar to the different characters of a drama. The inception is extremely laconic and clear. Further development reveals the exposed images and expands the relationship between them, which leads to tremoling multiphonics in growing up to fortissimo dynamics! It is the culmination of the first section, followed by another phase—quiet phase, as a response to or echo of distant dramatic events.

This is just one of the possible interpretations of the figurative circle of images and the semantic 'meaning' of this work. Everyone can hear something of their own in it, be it the sounds of nature, pictures of the universe, space, etc.

The notion of techniques as carriers of artistic images is an important concept. My composition *Quantum reality* for 12 performers (2020a) is based on creating multi-dimensional structures using contemporary composition techniques and extended methods of performance to musically depict phenomena of the quantum world (Fig. 17.8).

Coming back to the information theory in music, its another aspect is the question of 'what exactly music says?'. The fundamental question of meaning in music has been an immense field of research in philosophy, psychology, sociology of music, music cognition and music semantics with a tremendous amount of works produced. For example, more information about music semantics can be found in research of Philippe Schlenker (2017, 2022). We will not dwell into that territory. However, it is crucial to quote a phrase that Maestro Ivan Fedele said during his lectures at the Accademia Nazionale di Santa Cecilia in Rome, Italy (2017): 'Musica è la lingua autoreferenziale' (music is the auto-referential language). Given absolutely the highest imaginable level of abstractness of musical art, music itself always 'speaks' about itself (self/auto-referentiality).

Music is a language, and transfers information. But music is a language far more transcendental, sublime, abstract, elevated and exalted compared to ordinary human language and many other mediums. Unlike other arts (literary, poetic, theatrical and dramatic) that implement medium, music is a medium itself. Common understanding and speculation around things such as yes or no, white or black, good or evil, light or shadow are largely feeble in the territory of music.

The symbolic language of music is so powerful that music can simultaneously 'say' a lot of things, or have a lot of meanings. Just like electron can spin simultaneously in all directions, the language of music (especially avant-garde music) expresses all possible meanings at the same time. In short, music may say nothing but express everything! And this multi-vector poly-directional expressive power has been one of the primary reasons why music penetrates the very hidden corners of the human soul, which are otherwise inaccessible to everything and anything else.

Music always has/expresses/transmits a superposition of meanings. It is impossible to unambiguously say 'I write a book' or 'I read a book' in the language of music, but that's not the aim of the language of music. Music can express phenomena far beyond every imaginable capacity of human

Fig. 17.8 Beginning of the score of Rakhat-Bi Abdyssagin *Quantum reality* (2020a) (© Rakhat-Bi Abdyssagin (2020a). All Rights Reserved)

language. Music is simply on a different level compared to ordinary human language. There are many different human languages which need translation to be perceived, while music does not need any translation. Music can express what cannot be expressed by language (human conventional one). Every interpretation of music is only a conditional point of view, but not the music itself. It could be summed that there is a universal barrier between the *divine language*—music, and *human language*—language. The human language can only grasp the taste of divine (supreme) language—music, but cannot meaningfully replicate it or transmit without distortion.

There is no and can be no any single 'absolute' perception of music. Any quality of the perception or evaluation of music is strongly linked and depends on the level of development of the mind and the reason of the person who perceives or evaluates the music. This is especially relevant in the case of avant-garde music.

In June 2018 I was lucky to visit the CERN (European Organization for Nuclear Research) together with a delegation from Kazakhstan (Fig. 17.9). Scientific research is carried out at the very forefront at ultra-modern facilities such as the Large Hadron Collider (LHC). It was at CERN in 2012 that in the Large Hadron Collider a trace from the Higgs boson, called the 'God Particle', was discovered. For this, Peter Ware Higgs and François Englert were awarded the Nobel Prize in Physics in 2013.

Interesting to mention that Fabiola Gianotti, an Italian physicist and first female Director-General of CERN, in addition to her qualifications in

Fig. 17.9 Government delegation of the Republic of Kazakhstan to CERN, Geneva, June 2018

physics has also studied piano at the Conservatorio Giuseppe Verdi di Milano, Italy.

During this visit to CERN I was pleasantly impressed by the level of awareness of local scientists in avant-garde music. They not only know the names—Karlheinz Stockhausen, Iannis Xenakis, Luigi Nono, Luciano Berio, Olivier Messiaen among others—but are familiar with their styles. One of the physicists during discussion said: 'to be honest, I don't like Stockhausen's music, it hits me in the ears, but I cannot but recognise its greatness, because this music changed the world' (Abdyssagin 2020b, pp. 15–16). And here we note the fundamental aspect of the perception of avant-garde art. Many people speak and operate with the phrases 'I like it' or 'I don't like it'. But when it comes to high musical art, expressions such as 'I understand it' or 'I don't understand it' are more appropriate. Inspired by this visit to CERN I wrote a symphonic picture *The Sacred Universe of Particles* for large orchestra (2018) (Fig. 17.10). This orchestral work was premiered during my symphonic portrait concert *Ghosts of Immortality* in Great Hall of the Jambyl Philharmonie Almaty by the State Academic Symphony Orchestra of Kazakhstan on 28th August 2022.

Considering the 'like-or-not-like' formulations, it is reasonable to quote R. Feynman's illuminating commentary on that matter: '…if we have a set

Fig. 17.10 Excerpt from the score of Rakhat-Bi Abdyssagin *The Sacred Universe of Particles* (2018) (© Rakhat-Bi Abdyssagin (2018). All Rights Reserved)

of "strange" ideas, such as that time goes slower when one moves, and so forth, whether we *like* them or do *not* like them is an irrelevant question. The only relevant question is whether the ideas are consistent with what is found experimentally' (2011, p. 77).

References

Abdyssagin, R.-B. 2012. *Light through fog* for flute solo. Berlin: Verlag Neue Musik.

Abdyssagin, R.-B. 2013. *Mathematics and Contemporary Music*. Almaty: Kazak Universiteti.

Abdyssagin, R.-B. 2018. *The Sacred Universe of Particles* for symphony orchestra.

Abdyssagin, R.-B. 2019. *Serenata di stelle invisibili* per flauto, clarinetto, sassofono contralto, violoncello e pianoforte.

Abdyssagin, R.-B. 2020a. *Quantum reality* for 12 performers.

Abdyssagin, R.-B. 2020b. *Serenada nezrimyh zvezd* (Serenade of Invisible Stars). Almaty: Kazak Universiteti.

Abdyssagin, R.-B. 2022a. *Noveishie kompozitorskie i ispolnitel'skie tehniki* (The Newest Composing and Performing Techniques). Astana: KazNUA.

Abdyssagin, R.-B. 2022b. *Selected Solo Works* (anthology) (2012–2021). Berlin: Verlag Neue Musik.

Boole, G. 1854. *An Investigation of the Laws of Thought*. London: Walton & Maberly.

Feynman, R., R.B. Leighton, and M. Sands. 2011. *Six Not-So-Easy Pieces: Einstein's Relativity, Symmetry, and Space-Time*. New York: Basic Books.

Schlenker, P. 2017. Outline of Music Semantics. *Music Perception* 35 (1): 3–37.

Schlenker, P. 2022. Musical Meaning within Super Semantics. *Linguistics & Philosophy* 45: 795–872.

Vedral, V. 2018. *Decoding Reality: The Universe as Quantum Information*. Oxford: Oxford University Press.

18

Basic Operations with a Tone Row

This chapter will concisely observe and present different methods and formal systems of structuring pitch-relations in contemporary music, starting from the most basic and the simplest arithmetic operations with a tone row (foundations of the twelve-tone technique).

A tone row (series) was one of the most iconic phenomena in the first half of twentieth century music, the basis of system-forming directions such as dodecaphony, integral/total serialism and other compositional methods. In a certain sense serial techniques can be seen as a modern example of applying the mathematical apparatus to music. A tone row (series) is a central element of dodecaphonic composition, the center of its intonational resources. Serial music composition is based on repetition of the same tone row (selected individually for a given composition) in different versions, and the tone row is a sequence of all the available 12 tones of an octave (within Western equal temperament). Dodecaphony (twelve-tone technique) is the foundational theory for writing music for twelve equal and independent tones (Abdyssagin 2013, p. 11).

The main principle of the dodecaphony is that a tone (belonging to a tone row) cannot be repeated until after all other 11 tones pass. The sequence of 12 tones forms a tone row. For instance, a chromatic scale within an octave is a striking 'canonical' example of a tone row, exactly because a chromatic scale within one octave (according to the Western equal temperament) consists of 12 non-repeating independent and equal tones.

Understanding the essence of a tone row and being able to structure it and apply in a composition is an essential step in mastering the skills of operating with the twelve-tone technique. The key methods of varying and developing a

© The Author(s), under exclusive license to Springer Nature
Switzerland AG 2024
R.-B. Abdyssagin, *Quantum Mechanics and Avant-Garde Music*,
https://doi.org/10.1007/978-3-031-63161-0_18

tone row are: retrograde (also called 'cancrizans'), inversion, combination of these two methods as a 'retrograde inversion', rotation, mutation, etc. The foundational aspects of operating with a tone row are very similar to the methods and concepts (even a formal language) of medieval, renaissance and baroque polyphonic techniques. Many scholarly works have been devoted to the art of working with a tone row, notably, a series of books by composer Ernst Krenek (1937, 1940, 1958a, 1958b, 1959) is a good example.

The evolution of a tone row is inextricably linked with giants such as Arnold Schoenberg, Alban Berg, Anton Webern, Pierre Boulez, Karlheinz Stockhausen, Luigi Nono, Luciano Berio, Luigi Dallapiccola and others. Many composers who did not necessarily fully link their creativity with dodecaphony and/or serialism, in one way or another implemented tone rows in their works. Among them are Igor Stravinsky, Olivier Messiaen, Witold Lutosławski, Edison Denisov, as well as spectralist composers and those who work in the areas of sonoristics, nano-music, saturation etc.

I will show a clear example of basic operations with a tone row (Abdyssagin 2022, pp. 26–28). Here is my initial prime tone row (at the moment only the pitches, without durations, dynamics and other parameters, Fig. 18.1).

Rakhat-Bi Abdyssagin (2020)

Fig. 18.1   Initial prime tone row (only the pitches)

And now the same tone row is in retrograde (written backwards, Fig. 18.2).

Fig. 18.2   Tone row in retrograde

Here I add a rhythmic dimension to the initial prime tone row (Fig. 18.3).

Fig. 18.3   Tone row with durations

And its retrograde (including the rhythm, Fig. 18.4).

Fig. 18.4   Retrograde of the tone row with durations

It is also possible to make 'separated' retrogrades: writing only the rhythms (durations) backwards while pitches remain intact, or writing pitches backwards while preserving the original rhythmic sequence.

Here we see an example of inversion (Fig. 18.5) of the original tone row (when the same intervals go in the opposite direction compared to the prime tone row).

Fig. 18.5   Inversion of the tone row

Here comes the inversion of the retrograde (the retrograde of the original tone row undergoes inversion, Fig. 18.6).

Fig. 18.6   Inversion of the retrograde

And now the retrograde of the inversion (the inversion of the original tone row is retrograded, Fig. 18.7).

Fig. 18.7   Retrograde of the inversion

Let us consider the 'mutation' of the tone row. The original prime tone row is given once again (Fig. 18.8).

Fig. 18.8 Original prime tone row

The chromatic scale from the first tone of the prime tone row (Fig. 18.9).

Fig. 18.9 Chromatic scale from the first tone

Circle of the fourths and fifths from the first tone of the prime tone row (Fig. 18.10).

Fig. 18.10 Circle of the fourths and fifths from the first tone

Finally, a new tone row emerged (Fig. 18.11), 'woven' by the mutation of the original prime tone row through the chromatic scale and the circle of fourths and fifths.

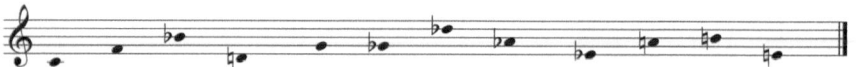

Fig. 18.11 A new tone row born out of the 'mutation' of the original one

How did this new tone row appear? Now I describe an example of the process of mutation of a tone row. The 1st tone C remains. The 2nd tone of the original prime tone row is H. In the chromatic scale H is the 12th tone. In the circle of fourths and fifths the 12th tone is F. Now we take F and write it after C in the new tone row.

Let us go further. The 3rd tone of the original prime tone row is B♭. B♭, or more precisely A♯ (B♭ and A♯ are the same pitches according to the Western equal 12 tone temperament, so present the enharmonic equivalence), is the 11th tone of the chromatic scale. The 11th tone of the circle of fourths and fifths is B♭, so we put B♭ after F as the 3rd tone of the new tone row.

Then, operating according to the same method, we write down all remaining tones and receive a new tone row as a result of mutation.

The art of working with a tone row is largely determined by the imagination and ingenuity of the composer, as well as his innovation. This topic is especially useful for beginning composers and students; it describes the basics of working with a tone row. This is an introduction to an incredibly large and broad topic. Mastering dodecaphony and serialism is not an easy task and requires deep knowledge and preparation. The further evolution of the serial technique, in particular, P. Boulez's 'total (integral) serialism' can be studied after mastering these basics. The methods described here are the simplest and most elementary operations possible, they present the basic arithmetic of the twelve-tone technique. These serial operations are similar to elementary numerical operations of arithmetic such as addition, subtraction, multiplication, and division.

Despite the intellectual and ideological magnitude, the main limitation of the twelve-tone technique is embodied in its basement on the equally tempered 12 tones of the octave. Obviously, the length and combinatorial nature of tone rows (series) vary, however, the 'classical' dodecaphony is valid only within the frames of 12 tones, while in reality, as we know, the spectrum of sounds is continuous, so you can 'divide' as much as you want the distance between the tones. As John Cage wrote: 'Curiously enough, the twelve-tone system has no zero in it. Given a series: 3, 5, 2, 7, 10, 8, 11, 9, 1, 6, 4, 12 and the plan of obtaining its inversion by numbers which when added to the corresponding ones of the original series will give 12, one obtains 9, 7, 10, 5, 2, 4, 1, 3, 11, 6, 8 and 12. For in this system 12 plus 12 equals 12. There is not enough of nothing in it' (2022, p. 79).

Alhough Cage speaks of zero, it is absolutely not that zero which I discuss when analysing my composition *Shadows of the Void* (2019) in Chapter 16. Cage's zero mentioned here has nothing to do with it. Cage refers to the 'closeness' of the systematic concept of the twelve-tone technique. Musically and practically any addition, multiplication and other 'increasing' operations are impossible and meaningless when a 12 tone row is taken as a basis of the systematic framework in case of Schoenberg's dodecaphony. Of course, arithmetically, composer can do almost anything imaginable with the digit of 12. But, as Cage displayed, within Schoenberg's dodecaphonic system the only result of the simplest of equations can be $12 + 12 = 12$ or $12 \times 2 = 12$ or $12 \times 12 = 12$ and so on.

In principle, the equation of $12 + 1 = 13$ is also correct for music, but within the twelve-tone system of equal temperament and within Schoenberg's dodecaphony the musical and practical sense or meaning of this equation is the same as of Eq. $12 + 1 = 1$ because the resulting note '13' will be just the same the note '1' performed one octave higher.

In the physical world, 'as far as we know today' (Feynman et al. 2011, p. 25), the highest existing speed (upper limit) of exchanging information (travel of matter or energy) through space is the speed of light (approximately

300,000 kms per second). It is a physical constant. As per today's knowledge, there is no speed higher than the speed of light. Therefore, even though you can mathematically add or multiply c (speed of light), in physical reality it is impossible and meaningless because the speed of light added or multiplied to the speed of light will always result in a (single) speed of light ($c + c = c$ or $c \times 2 = c$ or $c \times c = c$). Hence, another correlation, that in the relativistic world any additions/multiplications of the speed of light to itself will result in a (single) speed of light, while in music, in Schoenberg's dodecaphonic system, any additions/multiplications of 12 will result in 12!

The speed of light as a physical constant is crucial in A. Einstein's famous equation $E = mc^2$ and the relativistic worldview. In fact, only conventional arithmetic operations are not valid with the speed of light, while relativistic mechanics already has specific mathematical apparatus to address this issue.

The described analogy between the limitations of 12 in music and relativistic physics is very fragile and distant, but it can be listed among possible correlations.

Very important note is that all of the above mentioned operations with 12 are valid only and exclusively in the case of Schoenberg's dodecaphony, where the same notes in different octaves are considered as identical. It is obvious that already while dealing with structuralism and serialism of Pierre Boulez, Karlheinz Stockhausen, Luigi Nono and others the aforementioned principle completely loses its validity.

Additionally, the above operations are valid only within the context of a 12 tone equal temperament, and lose their validity for other types of temperaments. However, the principle itself can still be used with certain corresponding amendments, like, say, in 24 tone equal temperament there will be no limitation of 12, but there will be respective limitations of 24 (24 + 24 = 24)!

While certain aspects remain valid for any 'high art' music independently of its time, the conventional and traditional analytic apparatus as well as methods of approach mostly show their incapacities and inadequacy when dealing with contemporary music. Therefore, it seems reasonable that many techniques of new music require respectively new systems of analysis.

In terms of musical composition the essence of limitations reviewed above was brilliantly yet succinctly explained by Olivier Messiaen: 'This charm [of impossibilities], at once voluptuous and contemplative, resides particularly in certain mathematical impossibilities of the modal and rhythmic domains. Modes which cannot be transposed beyond a certain number of transpositions, because one always falls again into the same notes; rhythms which cannot be used in retrograde, because in such a case one finds the same order of values again—these are two striking impossibilities' (1956, 1 Volume: Text, p. 13). Additionally, O. Messiaen brilliantly represents the compositional 'apparatus' (or arithmetic calculations) of how composer can deal with these 'impossibilities' (1956, 2 Volume: Musical Examples).

It is important to understand that any new musical technique does not 'cancel' or make any preceding technique invalid; in contrast—if a new technique is valid per se, then it will 'include' in itself the previous ones as specific facets (given that preceding ones are also valid).

References

Abdyssagin, R.-B. 2013. *Mathematics and Contemporary Music*. Almaty: Kazak Universiteti.

Abdyssagin, R.-B. 2019. *Ombre del Vuoto* per 11 esecutori ('Shadows of the Void' for 11 performers).

Abdyssagin, R.-B. 2022. *Noveishie kompozitorskie i ispolnitel'skie tehniki* (The Newest Composing and Performing Techniques). Astana: KazNUA.

Cage, J. 2022. *Silence. Lectures and Writings*. London: Marion Boyars.

Feynman, R., R.B. Leighton, and M. Sands. 2011. *Six Not-So-Easy Pieces: Einstein's Relativity, Symmetry, and Space-Time*. New York: Basic Books.

Krenek, E. 1937. *Über neue Musik: Sechs Vorlesungen zur Einführung in die theoretischen Grundlagen*. Wien: Verlag der Ringbuchhandlung.

Krenek, E. 1940. *Studies in Counterpoint Based on the Twelve-tone Technique*. New York: G. Schirmer.

Krenek, E. 1958a. *Tonal Counterpoint in the Style of the Eighteenth Century*. New York: Boosey and Hawkes.

Krenek E. 1958b. *Zur Sprache gebracht. Essays über Musik*. Munich: A.Langen, G.Müller.

Krenek, E. 1959. *Modal counterpoint in the style of the sixteenth century*. New York: Boosey and Hawkes.

Messiaen, O. 1956. *The Technique of My Musical Language*, trans. J. Satterfield. Paris: Alphonse Leduc.

19

Tears of Silence

Apart from the traditional tonal system and harmony as well as the twelve-tone technique, another major approach to systematising interrelations between pitches is what Italian composers call *campo armonico*—the harmonic field (Abdyssagin 2020, pp. 38–42). On the one hand, it is a rather vague and wide definition of techniques and conceptions about the formal language of systematising pitches. On the other hand, the harmonic field is a powerful tool for designating and determining dominant features or even *forces* within a specific group of tones.

However, while groups of tones are mentioned for convenience of description, a harmonic field is not so much concerned with the tones themselves but rather with phenomena embodied and emitted by interrelations between the tones of a group. Hence the prime hallmarks of harmonic fields are radiated and absorbed by almost intangible musical links or sets of interrelations between tones or groups/clusters of tones.

The harmonic field is much wider than the conventional tonal system and allows recognition of far greater and deeper amount of aspects of 'family resemblance' (in quite a Wittgensteinian sense) between its members. Compared to the twelve-tone technique or integral/total serialism, the harmonic field is less pre-determined by its initial algorithms, and thus allows a composer to have a greater control over entities contained in the harmonic field and their combinations. At the same time, the broad nature of the harmonic field allows it to incorporate the conventional tonal system as well as the twelve-tone technique and serialism as specific domains within the paradigm of the harmonic field. Additionally, the harmonic field can be used in mutually enriching fusion with contemporary composition techniques,

R.-B. Abdyssagin, *Quantum Mechanics and Avant-Garde Music*, https://doi.org/10.1007/978-3-031-63161-0_19

thus producing the concepts of 'harmonic-sonoristic field' or 'harmonic timbral-textural field' and many other possibilities.

One of the immanent characteristics of a harmonic field and working with it is connected with what Maestro Ivan Fedele calls 'saper leggere il materiale'. This means 'know how to read the material'. It is a paramount skill that should to be developed on a way to true mastery of the art of composition. 'Saper leggere il materiale' essentially means analytic ability and proficiency in finding and developing all the possible correlations and potential kernels within even the tiniest and simplest musical elements.

Almost all great composers of the past were preeminent masters in 'saper leggere il materiale'. Johann Sebastian Bach and Ludwig van Beethoven are great historic examples as they constructed whole musical universes from an infinitesimal initial material and were able to find unexpected solutions and connections of even the most trivial musical elements and their combinations. In conjunction with fantastic and unimaginable cosmic initial 'building material', their exceptional supremacy of command of music's formal and semitic language unveiled the true geniuses of these composers. Among the giants of the twentieth century Franco Donatoni is usually noted for his exaggerated notion and application of 'saper leggere il materiale'.

To demonstrate some personal approaches of various possibilities for applying *campo armonico* in music, I conducted structural analysis (Abdyssagin 2022, pp. 103–105) of the very beginning of my Concerto for piano and symphony orchestra *Tears of Silence*, which was composed in 2018 (Fig. 19.1).

We can focus on the first five bars, before the piano part is played. It is also possible to say that after writing these first 5 bars, all the 'creative' part of the work was a kind of 'completed', and the rest can be considered as the fruit of the 'technical' approach in terms of composition methods. Moreover, this is one of the possible paths for describing the process of creating a piece of music: to find 'thematic' (fundamental) material, exhibit it, and then develop it to organise the musical form as a complete structure in order to preserve and maintain the general concept.

This might be demonstrated by a simple example. Let us look at the part of Flute 1 (Fig. 19.2), after which we turn our attention to the part of Oboe 1 (Fig. 19.3).

This is a kind of 'augmentation' of a fragment from the part of Flute 1.

Similarly the beginning tones—C, F, A, Ab, G—of the part of Oboe 2 (Fig. 19.4) are repeated in 'diminution' in the part of Flute 2 (Fig. 19.5).

Looking further (Fig. 19.6); the second note (which appears after the very first note F) on the part of Flute 1 is C♯, and the interval from C♯ is a minor

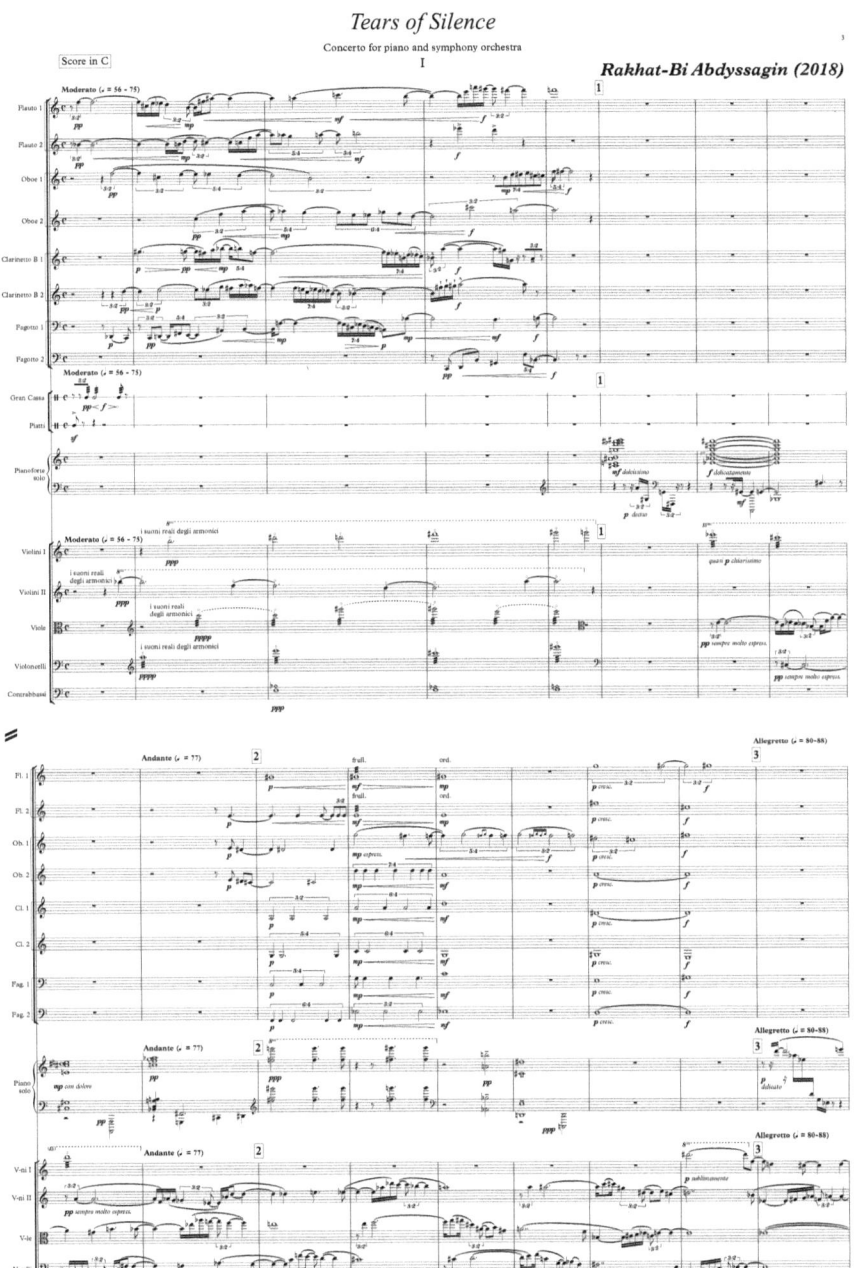

Fig. 19.1 First page from the score of Rakhat-Bi Abdyssagin *Tears of Silence* (2018) (© Rakhat-Bi Abdyssagin (2018). All Rights Reserved)

Fig. 19.2   Flute 1

Fig. 19.3   Oboe 1, fragment in 'augmentation'

Fig. 19.4   Oboe 2

Fig. 19.5   Flute 2, fragment in 'diminuition'

third up (from C♯ to E), while on Flute 2 the second note is H (appears after the first note B♭), and the first interval from H is also a minor third up (H to D). Then H, F♯, A♯ appear. From that point this chord 'decreases', goes down by half a tone, and then, from below goes half a tone up, and as a result, C, F, A are obtained.

We now pay attention to the tuplet (7:4) in the part of Clarinet 1 and Clarinet 2 and the Bassoon 1 (Fig. 19.7). In this case, the group of tuplets have the same notes, which are modified by the technique of interpolation and polarisation, sometimes simply changing the places of the pitch.

Next, let us consider the part of Violins I: H, D♯, D, F♯, G♯, G (Fig. 19.8).

The parts of Flute 1 and Oboe 2 end in F and F♯. And this is not an accidental occasion. Given that in the parts of Violins I and Violins II (considered together) the first interval is a minor second—B♭ and H—at the end of this

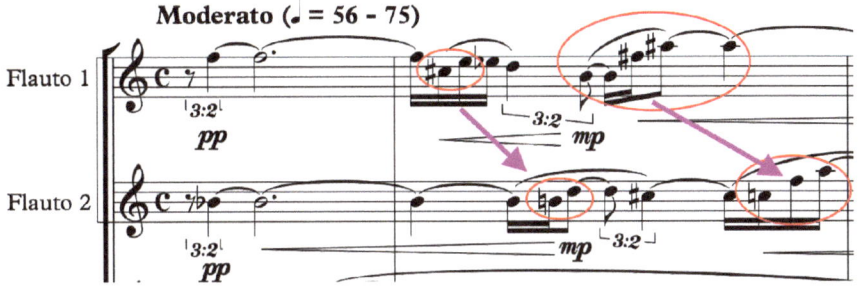

Fig. 19.6 Parts of Flutes 1 and 2 from Rakhat-Bi Abdyssagin *Tears of Silence* (2018)

Fig. 19.7 7:4 tuplets in the parts of Clarinet 1 and 2 and Bassoon 1

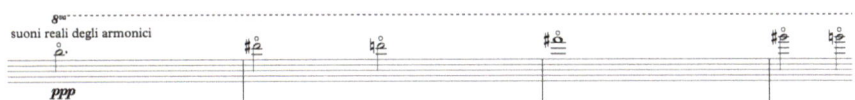

Fig. 19.8 Part of Violins I

block this minor second (in parts of Violins I and II) is represented in inversion as a major seventh – F and F♯ – in the parts of Flute 1 and Oboe 2.

Now we come to the first chord (Fig. 19.9) that appears in the part of solo piano and investigate how this chord was constructed (right at the beginning of Rehearsal Number 1, Fig. 19.10). We should remember that in the part of Violins I (pitch line with harmonics) the first sound is H and the last sound (within the frame of this nano-phase) is G.

H_4 and G_5 together make a minor sixth (inversion of a major third), which is obvious. This is the reason why a minor sixth—H_4 and G_5—make the central element of the first chord of the piano part.

Fig. 19.9 First chord played by piano solo

Fig. 19.10 Beginning of piano solo part in the Rehearsal Number 1 (including the first chord which is followed by descending minor ninths)

The first interval that appears in the score is the fifth—F_5 in part of Flute 1 and Bb_4 in part of Flute 2). In correlation with this, in the piano chord, from H_4 downwards E_4 is indicated (which gives the fifth). Further from E_4 downwards $C\sharp_4$ (minor third) is given; this correlates with the minor third interval ($C\sharp$ to E) that appears immediately after the first note F in the part of Flute 1. Within the first 5 bars the last interval between the parts of Flute 1 and Flute 2 is a major seventh (F_6 in Flute 1 and $F\sharp_5$ in Flute 2). There, a minor second (which is the inversion of a major seventh) is drawn from H_4. This explains how the tones of the lower range (below H_4) of the first piano chord were selected.

The tones that comprise the higher range of the first piano chord (above G_5) are D_6, $D\sharp_6$, $F\sharp_6$, $G\sharp_6$, which is a 'filtered' line of harmonics from the part of Violins I in the first 5 bars, with the difference that now they appear not 'horizontally' (as a 'melody'), but rather vertically (as a harmony).

During the sounding of this chord, on the left-hand staff the chromatic passages are indicated going down with an interval of a minor ninth (which is a minor second through an octave), forming tones A_4, $G\sharp_3$, G_2, $F\sharp_1$ (Fig. 19.11).

Later, this 'horizontal' line of the minor ninths was transformed into a 'vertical' one in the same piano part in the first bar of Rehearsal Number 2 (Fig. 19.12).

Fig. 19.11 Minor ninths

Fig. 19.12 'Horizontal' minor ninths became 'vertical' chords

Fig. 19.13 Piano part starting from the Rehearsal Number 4

This sequence (again in horizontal form, but already in minor seconds) will become an important thematic element in the piano part starting from the Rehearsal Number 4 (Fig. 19.13); with the addition of F at the bottom, since F is the first definite pitch that appears in the score (first bar in the part of Flute 1).

In the second bar of Rehearsal Number 1, Violas play the same pattern (Fig. 19.14) that was present in the part of Flute 1 at the beginning (Fig. 19.15), and the same pattern was also imitated by Violins II (Fig. 19.16). In both cases of imitation (by Violas and Violins II) the pattern begins from different starting tones, which adds to the harmonic diversity of the overall musical space and harmonic field.

Fig. 19.14 Imitation in part of Violas

Fig. 19.15 Original pattern by Flute 1

Fig. 19.16 Imitation in part of Violins II

Fig. 19.17 Left hand of piano solo

The element in the upper voice in the left hand of piano solo (Fig. 19.17)—G♯, G, F♯—is a 'hint' to the future reference point in the piano solo part on one hand, and on the other part is 'predicts' what the oboe will repeat a minor second lower in the second bar of the Rehearsal Number 2 and beyond by playing G, F♯ and F (Fig. 19.18).

Fig. 19.18 Oboe 'echoes' elements from the piano part

Now we made a structural analysis of certain aspects of only 5 initial bars of this piano concerto and correlations between these 5 bars and the rest of material in the first two pages of the score. We have not analysed the whole concerto at all, but only the 5 initial bars and the first 2 pages of the score, and even then what we investigated and revealed is only 1/10 (one tenth) of what is present in these 5 initial bars and first two pages. This is not a full analysis of the first two pages of the score, nevertheless, even what we have just described already elucidates the main correlational powers of the phenomenon of the harmonic field and introduces the cogent and coherent method of structural analysis. Different elements of this method of analysis have existed in composition, musicology and music theory for a very long time, however, the key feature is developing the ability (skill) to find all possible and feasible correlations and interconnections between separate entities present in the analysed material (Fig. 19.19).

As discussed, such logic and deduction (with 'saper leggere il materiale') provide powerful and appropriate tools for conducting structural analysis of this concert. By applying this method, more and more subtle interweaving of pitch, rhythmic and dynamic lines, as well as their embodiment in the orchestral fabric can be found. When I was composing this concert, all of that happened intuitively. I did not consciously think about these correlations and other structural aspects. These things happen automatically, almost at the level of reflexes. Like grammar or spelling which we learn in childhood; when we write a sentence, we do not consciously think (or realise) that we think about grammar, but we still apply it because we acquired it and it is present at the sub-conscious level.

Fig. 19.19 Second page from the score of Rakhat-Bi Abdyssagin *Tears of Silence* (2018) (© Rakhat-Bi Abdyssagin (2018). All Rights Reserved)

References

Abdyssagin, R.-B. 2018. *Tears of Silence*, Concerto for Piano and Symphony Orchestra.

Abdyssagin, R.-B. 2020. Noveishie kompozitorskie tehniki i genealogiya sovremennoi muzyki (The Newest Composition Techniques and Contemporary Music Genealogy). *Journal of Philosophy, Culture and Political Science* 4 (74): 34–47. Almaty: Al-Farabi Kazakh National University.

Abdyssagin, R.-B. 2022. *Noveishie kompozitorskie i ispolnitel'skie tehniki* (The Newest Composing and Performing Techniques). Astana: KazNUA.

20

The Newest Performing Techniques

The rapid development of the newest composing techniques in the twentieth century triggered the development of extended methods of playing instruments. The emergence of advanced performing techniques in the middle of the twentieth century is inextricably linked with the establishment and evolution of timbre-texture co-ordinate which became one of the exceptional phenomena of new music.

The merging of timbre and texture and the formation of a new timbre-texture co-ordinate is an unprecedented phenomenon of avant-garde music. The timbre-texture co-ordinate itself began to curve spacetime in musical works, and numerous new techniques, styles and directions appeared because of it. In particular, sonoristics—one of the most promising compositional approachess—is also a vivid manifestation of timbre-texture co-ordinate. Additionally, the 'musique concrete instrumentale' and 'noise music' of Helmut Lachenmann and the avant-garde of Krzysztof Penderecki as well as other techniques greatly contributed to new directions.

The new paradigm of thinking of composers puts new demands on performers. The old 'established' performing techniques were no longer enough to express and reflect a completely new worldview and outlook of composers. This led to the emergence of new and expanded techniques for playing instruments. First of all, breakthrough happened in the techniques of playing woodwind instruments. Until the beginning of the twentieth century, it was believed that flute, oboe, clarinet, bassoon and their versions could only act as solo instruments that needed accompaniment. Now the attitude towards them has transformed to capture their more multifaceted and independent perception. It is sometimes difficult to find a contemporary

R.-B. Abdyssagin, *Quantum Mechanics and Avant-Garde Music*, https://doi.org/10.1007/978-3-031-63161-0_20

work that does not, in one way or another, use new techniques for playing instruments.

The style and musical language of such iconic composers as Beat Furrer, Jose Manuel Lopez Lopez, Michael Jarrell, Toshio Hosokawa, Kaija Saariaho, etc., cannot be imagined without new instrumental techniques.

Considering some musical examples, the Italian composer Pierluigi Billone's cycle *Legno Edre I-V* (2003–2004) for solo bassoon can serve as a demonstration of the rebirth of this instrument in a completely new sonorous quality (Fig. 20.1). By developing timbral ideas within the framework of large-scale dramaturgy, this opus becomes a precedent of 'sonorous' symphonism in the projection of one instrument, greatly enhanced by a variety of extended techniques.

Now let us review some of the most significant instrumental techniques (Abdyssagin 2022a, pp. 30–35):

The *microtone* technique (also referred to as the *quarter-tone* technique) is one of the unique aspects of the perception and implementation of modern musical concepts. This is an example of further variation in the discretisation of tones and moving closer to the original, acoustically continuous spectrum of sounds. Since the middle of the twentieth century, microtonal technique began to dominate and prevail in a number of works and even areas of new music. The existence of microtones has been known since ancient times; however, their meaningful and structured use in the Western European tradition began only during the twentieth century. Microtone-based music is also sometimes referred to as 'microtonality', 'micro-polyphony' and even 'nano music'. One of the most influential composers contributing to the field of microtonality and micro-polyphony is Georg Friedrich Haas.

If historically it was believed that it was impossible to play chords on a flute, oboe, clarinet or bassoon—that is, they could play, relatively speaking, one note, and were monophonous instruments—now the phenomenon of *multiphonics* is widely known and used.

Multiphonic is a sound phenomenon that occurs on various instruments. Most often this applies to woodwinds, rarely to brass, and in very rare cases multiphonics occur on string instruments (but that is another topic). Multiphonics on woodwind instruments are chords (clusters of tones sounding together simultaneously) of different densities and different intensities that are played on a given instrument. For example, woodwind multiphonic is when on a flute or oboe, a clarinet or bassoon and their versions, two or more sounds are performed simultaneously. Multiphonics can be in the form of double harmonics, as, for example, when two notes are played very quietly on a clarinet or oboe, and they sound so perfect that an acoustic illusion arises

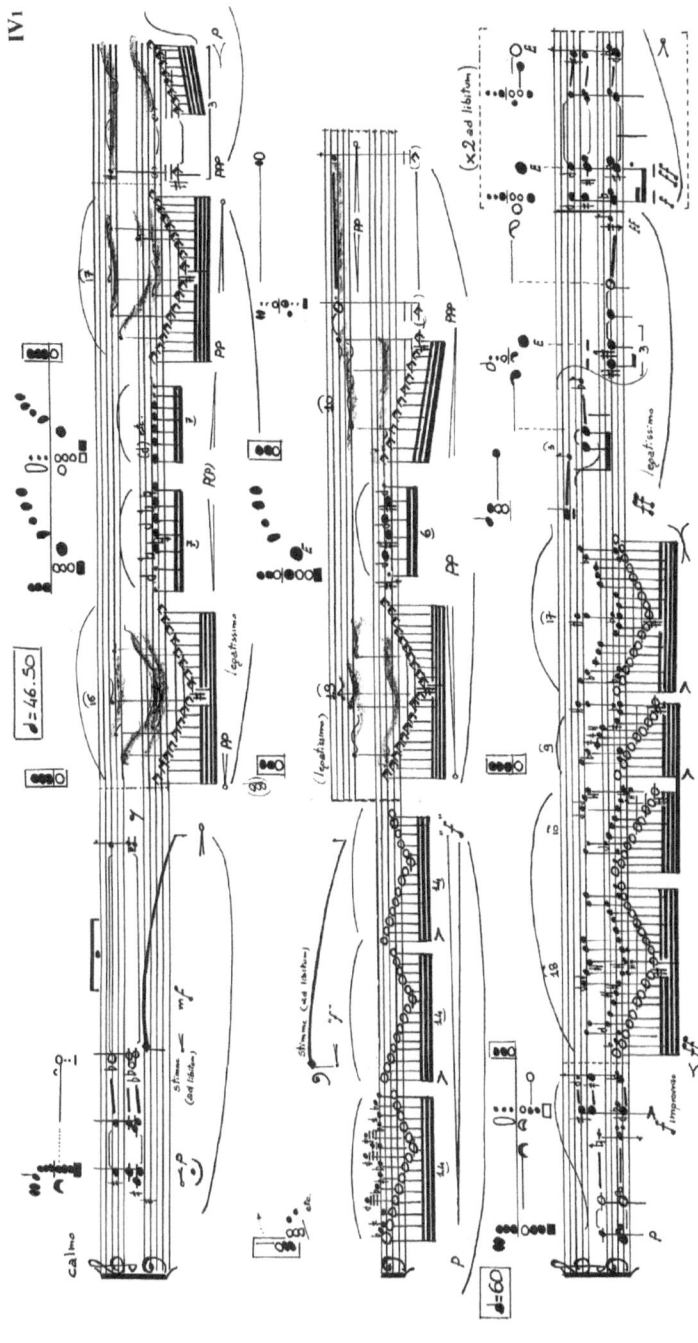

Fig. 20.1 Excerpt from the score of Pierluigi Billone *Legno Edre IV—Manda* (2003) (© *Pierluigi Billone*)

as if two clarinets or two oboes are playing simultaneously. Peter Veale's book (1994) gives a list of these two-voice multiphonics on the oboe, with detailed indications of the conditions for their playing (often quiet dynamics, middle register).

There are also types of multiphonics that consist of a large number of sounds, where sometimes it is even impossible to isolate (just by hearing) individual pitches in this huge, supermassive sound cluster. There are very flexible multiphonics with which you can play and combine *frullato* (flutter-tonguing) techniques, perform *bisbigliando* between multiphonics, *tremolo* between multiphonics, play them using the double tongue, triple tongue technique, staccato, play from *pp* to *ff*—this variety is considered the most flexible. On the other hand, there are also highly inflexible multiphonics that are much more difficult to perform and performing them carries an array of different restrictions (Abdyssagin 2017).

The spectrum of sounds of a number of multiphonics may correspond (albeit distantly, sometimes in the form of allegories) to certain harmonic structures, such as a C major chord or pure octaves, fifths and other intervals. Multiphonics for flutes with similar 'harmonic' properties are (to a limited extent) indicated in the fundamental treatise by Pierre-Yves Artaud (1995).

To competently use multiphonics in music works it is very important for composers to know which multiphonics should be used, of course, in addition to being aware of the fundamental principles of creating a work of art and evolving dramaturgy within the reality of new music. In the last century, several practical works have appeared on this topic. Starting from historical Bruno Bartolozzis *New Sounds for Woodwind* (1967) to cutting-edge publications. It is of critical importance to pay attention to verified sources, because it happens that, for example, a composer includes in his score a multiphonic from one book, but it cannot be used because it is impossible to play it, or the indicated fingering is incorrect. To properly use a multiphonic and make it performable, a composer must indicate the complete fingering for the performer in the score, sometimes even creamy indicating which finger is to press which key. The pressure of the air, breath, position of the embouchure etc. may also be indicated in the score.

As flutists say, on a flute only 20% of the air is used to generate sound, while 80% 'fly away'. On the other hand, on the oboe (considering the double reed and a number of other design features), almost 100% of the air is transformed into sound. This is the fundamental difference between these wind instruments and these features form the very basis of many advanced playing techniques.

Due to these features, the most stable multiphonics appear on bassoon and oboe. Less stable, but very interesting combinations of multiphonics are possible on the flute, for example when a tremolo is performed between the multiphonics. Not only may a sensation of tremolo arise, but at the same time a sensation of some kind of non-discrete continuous movement in parallel realities can occur due to the phenomenon of sound formation.

Considering the clarinets, it is important to acknowledge another feature of this instrument—there are clarinets of different systems. This is true of the majority of woodwind instruments, but in the case of clarinets it becomes saliently conspicuous. The most common clarinet system is the French system, nevertheless, there are also German and Italian systems. If classical repertoires can be played on the clarinets of different systems in approximately the same way (and produce roughly the same result), in regard to contemporary music—especially extended techniques—a lot starts to depend on a specific system and particular instrument. There is an excellent textbook about the clarinet written by Giuseppe Garbarino (1978), but it covers the clarinet of the Italian system. There is an excellent collection of multiphonics by Gerhard Krassnitzer (2002), but it includes multiphonics only for the German system clarinet.

The most extensive and foundational work on new sounds on flute was produced by the outstanding modern flautist Pierre-Yves Artaud (1995). I am pleased to note that we have been fruitfully collaborating with maestro P.-Y. Artaud: for example, he wrote a foreword for the publication of my solo flute piece *Omaggio a Ivan Fedele* (2018a) in Germany, and his unique and exceptional flute orchestra—Orchestre de Flûtes Français (which consists of 24 flutes!)—commissioned and performed the world premiere of my composition *Des pensées de loin* pour 24 flûtes (2018b) (4 Piccolos, 6 Grandes Flûtes, 6 Flûtes altos en sol, 6 Flûtes basses, 2 Flûtes octobasses) in Paris in 2018, and made its studio recording and second performance in Opéra de Massy in 2022. Fundamental research on extended techniques and multiphonics was done for oboe by Peter Veale (1994), for the bassoon by Sergio Penazzi (1972) and Pascal Gallois (2009), and for trombone by Benny Sluchin (1995) and Yuri Kasparov (2020).

In addition to microtones and multiphonics, the technique of *bisbigliando* should be noted among other important extended performing methods. Bisbigliando (means 'whispering' in Italian) is when approximately the same pitch is performed with different fingerings, producing variations of timbre (of the same pitch); e.g. the same note sounds in different (sometimes radically) colors. The produced pitches are usually not 100% the same, but are as close as possible to each other. According to the very nature of this technique, starting from a certain overtone (in some cases this is a specific note or a specific register of the instrument) more and more fingerings (timbres/colours) to play the same note appear. Therefore, the generalised rule can be that the bisbigliando technique sounds best in the upper register, where the largest number of harmonics is present, which corresponds to an increase in different possibilities for performing a certain pitch. The main condition for the implementation of this technique is the presence of different fingerings for performing the specified note.

Techniques such as *vibrato* and *smorzato* have long been used by performers (consciously or not) as an effective means of musicality and expression. Since the beginning of the twentieth century, a number of composers began to use these techniques in a more meaningful way, basing and building on them the 'structural frame' of their music works. The 'status' and significance of such techniques can sometimes rise to the level of the sonorous and timbral foundation of music. There are many options for performing a vibrato when, say, dynamics, or pitch amplitude, or several parameters fluctuate at once. Smorzato can be explained as a rhythmic vibrato—a kind of sound impulses.

Frullato as a common technique manifested itself back in the nineteenth century. Now its combination with other techniques becomes popular, for example, when frullato is combined with multiphonics, microtones and/or air noise. Varieties of frullato such as *hrullato* (a more guttural sound) and *growl* can become carriers of vivid images in semantics of a piece of music.

The air noise technique is when a performer varies what percentage of sound and what percentage of air noise comes from the instrument during the performance. Air noise (and related effects) has become one of the fundamental techniques around which other parameters are layered in the music of many composers. The most common ways to perform air noise (sometimes referred to as 'soffio' in Italian) are 'full air noise' or 'half tone / half noise' (they can be phrased differently). These and other characteristics of air noise directly depend on the features of the construction and design of different instruments. The capabilities of each wind instrument should be specifically studied if a composer wants to reasonably and professionally use them in music. As a rule, the air noise technique works best in the lower range of wind instruments. In certain cases, the performance of air noise (with a fixed pitch) is possible only and exclusively in the lower register. The role of air noise in contemporary composition was researched by Martin Loridan in his PhD thesis *Souffle: Air and Breath as a Composition Material* (2022) at the University of Leeds.

Slap tongue, tongue ram, key clicks represent a range of innovative percussive techniques.

The *slap tongue* (also called *pizzicato* on wind instruments) is an energetic, sharp and dry slap of the tongue, a technique that derives from jazz and is common among guitarists. Slap tongue (sometimes abbreviated as 'slap.' or 'pizz.') sounds especially impressive on the flute and clarinet, and can sometimes be performed on **mf** dynamics. Despite this, in general this technique (and this category of techniques) can be better and more suitable categorised as 'quiet techniques', and in the vast majority of cases they are performable only and exclusively in the lower register. In some situations, in relation to these percussive effects, the rule can be applied that the lower the range is, the better, clearer and more effective percussive sound will be produced. Variations of slap such as flap and pizzicato can be played on bassoon. The differences between them are described in detail in the book by Pascal Gallois (2009, pp. 45–48).

Tongue ram is another percussive effect, mostly used in flute performance. It even expands the range of flute in a certain way. When it is performed, the produced tone sounds a major seventh lower than what is written (Tantsov 2011, p. 16). That is, when performing tongue ram technique, if the main tone (written) is C_4, then the produced tone will be $D\flat_3$. Or if the main written tone is $C\sharp_4$, then the resulting tone will be D_3. In terms of sound and acoustic properties, the produced effect of the tongue ram is very similar to that of the slap tongue. These techniques have naturally related sonoristic properties.

Key clicks (or *key noise*)—is the knocking of keys (without air flow), as the name suggests. The performer plays on the keys without blowing into the instrument. The result is a dry, quiet and mystical timbre of the keys themselves, with only a resemblance of a pitch.

The *jet whistle* is a powerful, abrupt and whistling air impulse. The effect is loud in nature, and vaguely reminiscent of the sound of a whistle.

Whistle tones is beautiful, exquisite, harmonically rich, quiet and elegant technique, built on the harmonic scale of a given tone (the base tone on which the *whistle tones* effect is performed).

There are, of course, many other important techniques that can be found in contemporary music. To get a complete understanding of these techniques, it is important to have auditory experience, that is, to listen to their sound. The progress and evolution of contemporary performing arts tend to be in a constant search, and it is quite possible that in the future the list of these techniques will be greatly expanded.

Each of these new techniques creates a completely new world in which modern composers navigate. While these extended methods expand the capabilities and 'set of tools' of a composer, it is crucial to be able to work with them very carefully because the knowledge of new techniques alone (just in itself) will not automatically rise a composer to the degree of a true master. In fact, what elevates a composer to the first-class master in terms of skills and ability to build dramaturgy—is the way a composer knows how to structure the elements of his work, even regardless of the initial 'building material': be it sonoristics, new techniques or something else. When writing music (and in terms of 'musical material'), it is of essential and existential significance for composers to know/realise what they are working with and the nature of what they are actually doing.

More detailed and exact explanations of the nature, types, essence and range of applicability of multiphonics and many other extended performing techniques such as microtones, bisbigliando, vibrato, smorzato, oscillato, frullato, hrullato, growl, air noise, half tones/half noise, airy tone, slap tongue, tongue ram, key clicks, jet whistle, whistle tones and the like and concrete/specific examples of their implementation are given in my textbook (2022a). Additionally, this textbook provides structural analyses of selected contemporary music works as well as of some of my own compositions, and these analyses directly display how extended techniques can be used to construct the spacetime of new music.

In fact, extended techniques are an inseparable part of musical language of many contemporary composers. Personally, I actively use them in my own music. For example, in 2022 my *Selected Solo Works (Sammelband)* has been published by Verlag Neue Musik Berlin (Germany). This anthology includes 21 titles (around 40 pieces)—specially selected works (among many others in all different genres) written in 10 years (2012–2021) for each instrument solo of symphony orchestra (Abdyssagin 2022b). More than half of the pieces comprising this anthology contain extended techniques as architectonical basis of compositional structure. According to experts, composing such an anthology is an extremely rare case, and as a previous example the cycle of *Sequenze* (1958–2002) by the great Italian composer Luciano Berio comes to mind.

I would like to illustrate (Abdyssagin 2022a, pp. 77–79) the use and thorough application of some extended techniques flute playing by example of my piece *Omaggio a Ivan Fedele* for solo flute (Fig. 20.2). The story of creating this work is interesting in its own way. While studying *Corso di Perfezionamento* (analogue of artistic doctorate) in composition at the Accademia Nazionale di Santa Cecilia in Rome, at the end of November 2018 Maestro

Ivan Fedele gave a special assignment to our group. Maestro Fedele introduced a formula—A; A + B; B; B + C—that can be used to build and define the structure of the work. It is a simple scheme from the point of view of dramaturgy: first element A is introduced, that is, the first phase; then phase A + B comes, which essentially is element A where element B is interspersed; then this is followed by 'pure' element B (without A), and finally element B + C appears.

In *Omaggio a Ivan Fedele* 100% identical recapitulation is not used, so direct repetitions are avoided. All repeated elements (and their own subelements) sound in a varied form, however, retaining their main features and structural/formal *gestalt*. In addition to simply following this formula (A; A + B; B; B + C) for the development of dramaturgy, this piece is inextricably linked with sonorous effects which appear not only as carriers of images, but also as the 'backbone' of the architectonics of the entire compositional structure.

The 'building material' of this piece is closely and deeply interconnected with many of the newest performing techniques; these and several other sonoristic effects act like the characters of a drama (Abdyssagin 2020, p. 39). Here we can briefly concentrate on the introductory phase of this score. In the beginning, *100% air noise* is given at a certain frequency (pitch). Then *100% air noise* transforms into '*50% air–50% sound*'. After that, the ordinary playing (ord.) begins, however, with extensive use of microtones (quarter-tones) combined with *bisbigliando* technique: this creates a special interrelation. Then the part of '*50% air–50% sound*' comes, with the addition of frullato (flutter-tonguing); later the *air noise* returns. Then the *slap tongue* technique is introduced: it is performed on the pitches which were defined by the original harmonic field of the composition. After that, the multiphonic is given (fingering from Artaud 1995). The use of multiphonics fits into the canvas of the 'drama'. In the case of this work, dramaturgical development and emerging structures are inextricably linked with the sonoristic effects as well as extended techniques, which are not only carriers of images ('imaginary objects') but they are also the 'backbone' of the architectonics of the entire composition.

The unique technique deserves a special attention: a tremolo between multiphonics. In this particular case, this technique is combined with a gradual transition from *air noise* (full air noise) to sound (ordinary sound) and to *key clicks*. This technique does not sound like a real/proper 'tremolo' but is more of a 'timbral event' that is in constant motion and permanent change. Due to unique vibrations, this combination of techniques leads to the

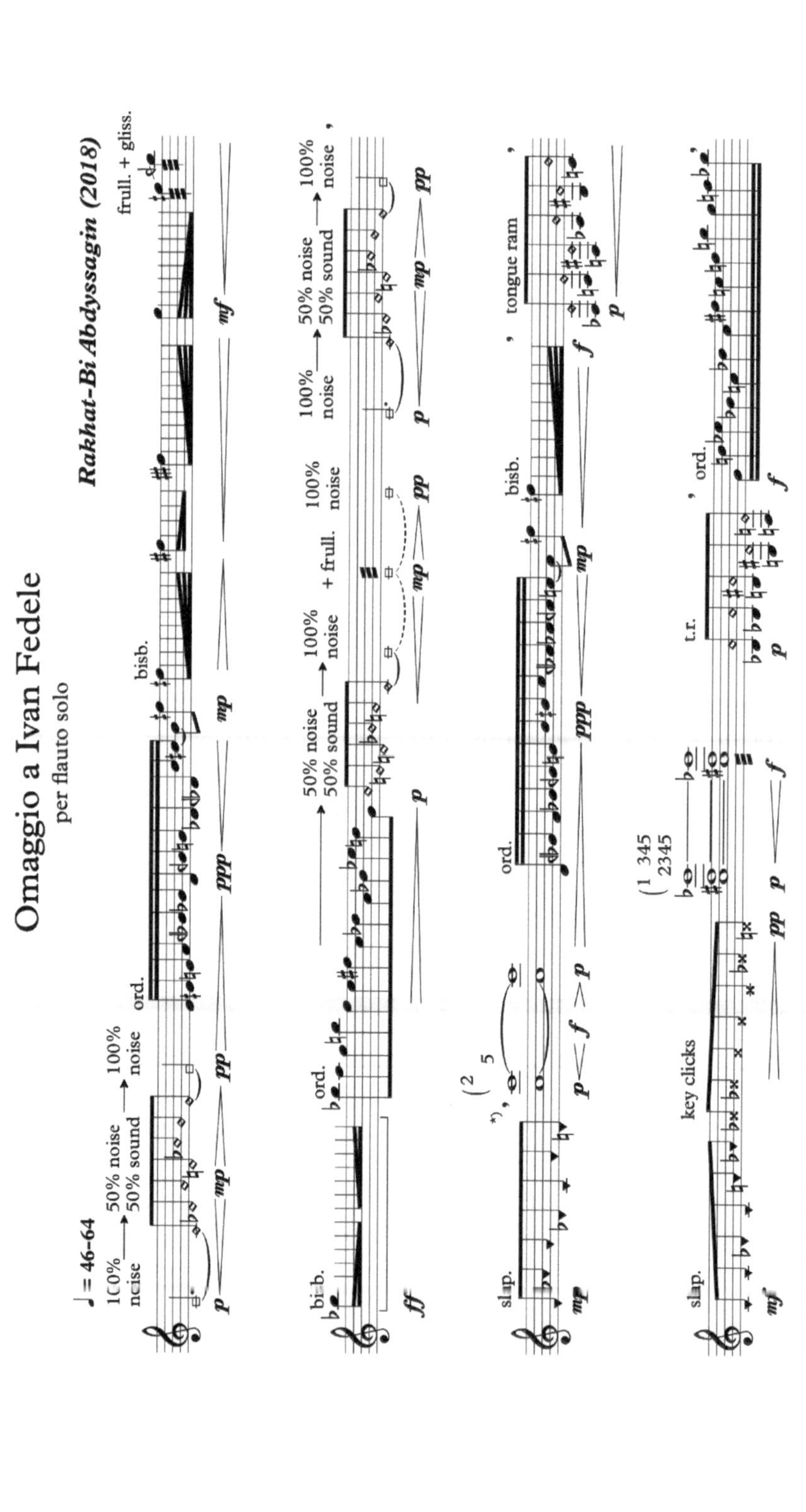

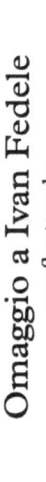

Fig. 20.2 Excerpt from the score of Rakhat-Bi Abdyssagin *Omaggio a Ivan Fedele* (2018a), published by Verlag Neue Musik Berlin

*) every ' indicates a few seconds of pausa

achievement of the magical effect of a sparkling sound stream, of a sonorous wave.

This piece is an illustration of how music works of various scales and for different instrumental combinations can be created with the help of some obvious and elementary formulas of a drama. In addition, the analysed composition is very convenient in terms of performance, since the techniques used are 'natural' and are closely related to the structural features and timbre of the flute. This is an example of the combinatorial nature of both extended methods for playing instruments and new compositional techniques in the form of a given formula for building a musical form.

References

Abdyssagin, R.-B. 2017. O novyh tehnikah igry na derevyannyh duhovyh instrumentah (On New Techniques of Playing Woodwind Instruments), pp. 100–107. Astana: KazNUA.

Abdyssagin, R.-B. 2018a. *Omaggio a Ivan Fedele* per flauto solo. Berlin: Verlag Neue Musik.

Abdyssagin, R.-B. 2018b. *Des pensées de loin* pour 24 flûtes.

Abdyssagin, R.-B. 2020. Noveishie kompozitorskie tehniki i genealogiya sovremennoi muzyki (The Newest Composition Techniques and Contemporary Music Genealogy). *Journal of Philosophy, Culture and Political Science* 4 (74): 34–47. Al-Farabi Kazakh National University.

Abdyssagin, R.-B. 2022a. *Noveishie kompozitorskie i ispolnitel'skie tehniki* (The Newest Composing and Performing Techniques). Astana: KazNUA.

Abdyssagin, R.-B. 2022b. *Selected Solo Works* (anthology) (2012-2021). Berlin: Verlag Neue Musik.

Artaud, P.-Y. 1995. *Flûtes au présent/Present Day Flutes*. Paris: Billaudot.

Bartolozzi, B. 1967. *New Sounds for Woodwind*. London: Oxford University Press.

Berio, L. 1958–2002. *Sequenze I-XIV*. Milan: Edizioni Suvini Zerboni, Vienna: Universal Edition.

Billone, P. 2003. *Legno Edre IV – Manda*.

Gallois, P. 2009. *The Techniques of Bassoon Playing / Die Spieltechnik des Fagotts/La technique du basson*. Kassel: Bärenreiter-Verlag.

Garbarino, G. 1978. *Metodo per Clarinetto*. Milano: Edizioni Suvini Zerboni.

Kasparov, Y. 2020. *Trombon. Evoliutsiya v XX veke i novye priemy igry* (Trombone. Evolution in XX Century and New Performing Methods). Moscow: Moscow State Tchaikovsky Conservatory.

Krassnitzer, G. 2002. *Multiphonics für Klarinette mit deutschem System und andere zeitgenössische Spieltechniken*. Aachen: Edition Ebenos.

Loridan, M. 2022. *Souffle: Air and Breath as a Composition Material*. PhD Thesis, the University of Leeds.

Penazzi, S. 1972. *Metodo per Fagotto*. Milano: Edizioni Suvini Zerboni.

Sluchin, B. 1995. *Practical Introduction To Contemporary Trombone Techniques*. Paris: Editions Musicales Européennes.

Tantsov, O. (2011) *Novye priemy igry na fleite* (New Methods of Playing Flute). Moscow: Moscow State Tchaikovsky Conservatory.

Veale, P., C.S. Mahnkopf, W. Motz, and T. Hummel. 1994. *Die Spieltechnik der Oboe/The Techniques of Oboe Playing/La technique du hautbois*. Kassel: Bärenreiter-Verlag.

21

Quantum Entanglement in Physics and Extended Techniques in Music

Quantum entanglement is one of the cornerstones of the current research in quantum physics. It is a specific phenomenon which depicts that when particles interact their quantum states cannot be measured independently and separately for each particle. Quantum entanglement is one of the key differences that distinguish quantum mechanics from classical mechanics, as entanglement is only present in quantum mechanics. For research on quantum entanglement, Alain Aspect, John Clauser, and Anton Zeilinger were awarded the Nobel Prize in Physics in 2022.

The story of quantum entanglement is very interesting. It started in 1935 with the publication of a paper by Albert Einstein, Boris Podolosky and Nathan Rosen, who conducted a thought experiment later known as the Einstein–Podolsky–Rosen (EPR) paradox. However, the main reason behind their publication was to show that 'the quantum–mechanical description of physical reality given by wave functions is not complete', with emphasis on local realism. The EPR paper received strong opposition in the response of Niels Bohr who stated that EPR's conclusion was fallacious, and made a remark that 'They [EPR] do it smartly, but what counts is to do it right' (Rae 2018, p. 51). N. Bohr also formulated his response as an article (1935) under the same title used by the EPR and published in the next volume of the same journal. In fact, quantum entanglement is older than the 1935 EPR paper; it is discussed in John von Neumann's book *Mathematical Foundations of Quantum Mechanics* (1932) and may even have appeared earlier. Nevertheless, EPR definitely made quantum entanglement famous.

The EPR paradox was not only the beginning of quantum entanglement but also one of the highest peaks of debate on the matter of quantum

R.-B. Abdyssagin, *Quantum Mechanics and Avant-Garde Music*, https://doi.org/10.1007/978-3-031-63161-0_21

mechanics. It is not reasonable to dwell deep into the story of that debate and EPR paradox, as this topic has already been widely covered, and here it is mentioned only as a succinct information about the emergence of quantum entanglement. Currently it is believed that the EPR paradox was mostly a critique of the Copenhagen Interpretation of quantum mechanics, and it does not violate quantum mechanics itself. The key point of the debate and source of the EPR paradox is that measurement in quantum mechanics radically differs from measurement in classical mechanics, which finally leads to the renowned measurement problem. The key argument made by EPR is that if quantum mechanics is a complete theory (no hidden variables), as in the Copenhagen Interpretation, then it must predict instantaneous action-at-a-distance. It was not the first time Einstein made this argument, but it was the first he used entangled systems.

The next step was Erwin Schrödinger's response to the EPR paradox in 1935, first in a letter to Einstein, where Schrödinger coined the term entanglement (originally Schrödinger used the German word *Verschränkung* and himself translated it as *entanglement*), and then in articles *Discussion of probability relations between separated systems* (1935a) and *Probability relations between separated systems* (1936). Interestingly, the famous Schrödinger's cat thought experiment also was devised in the same year 1935 and published in the article *Die gegenwärtige Situation in der Quantenmechanik* (The present situation in quantum mechanics) (1935b). This thought experiment was also aimed at demonstrating problems of Copenhagen Interpretation of quantum mechanics as well as touching the questions of understanding quantum superposition and reality itself.

Quantum entanglement plays a particular role in quantum information science and has given rise to many more fascinating quantum phenomena, such as quantum teleportation (teleportation of a quantum state), demonstrated in the foundational work of physicists Charles Henry Bennett, Gilles Brassard, Claude Crépeau, Richard Jozsa, Asher Peres, William Kent Wootters (1993), the no-cloning theorem (Wootters and Zurek 1982; Dieks 1982), the no-broadcasting theorem (Barnum et al. 1996), quantum cryptography (Ekert 1991), quantum computation (Deutsch-Jozsa Algorithm 1992, and P. Shor's algorithm 1994) and many other astonishing new fields of quantum information science. As Timpson stated (2013, p. 55), the field of quantum computation began 'with Deutsch's introduction of the concept of the universal quantum computer' in the article *Quantum theory, the Church-Turing principle and the universal quantum computer* (1985).

In the entangled state the whole contains more information than the sum of the parts. As Chiara Marletto explains, 'Entanglement arises when you have two or more quantum entities interacting; for example, two photons, or an electron and a photon. The essential feature of entangled quantum systems is that the information one can gain by jointly observing the systems is *more* than the information obtained by observing each system separately. […] So, when the qubits are entangled, it is *possible* to extract information *globally* (acting on both qubits) but *impossible* to do so *locally* (acting on each qubit separately)' (2021, pp. 125–127).

Analogous to this, in avant-garde music there are sonoristic techniques where sounds are also almost somehow 'entangled' or exist in the form of inseparable combinations. This became possible because of the emergence and evolution of the extended techniques of performing instruments. In particular, this revolution was apparent in woodwind instruments, and numerous important contributions to knowledge were made by great instrumentalists who wrote books elucidating and providing a contextual fundamental background on the usage of new performing techniques. Namely, flute treatise by Pierre-Yves Artaud (1995), oboe book by Peter Veale (1994), clarinet book by Phillip Rehfeldt (2003), bass clarinet multiphonics chart by Sarah Watts (2015), bassoon book by Pascal Gallois (2009), saxophone multiphonics by Daniel Kientzy (1989) and Marcus Weiss & Giorgio Netti (2010) and so on.

This is an excerpt from my piece *Omaggio a Ivan Fedele* for flute solo (2018a, 2022), where a very specific effect (tremolos of multiphonics) is present (Fig. 21.1).

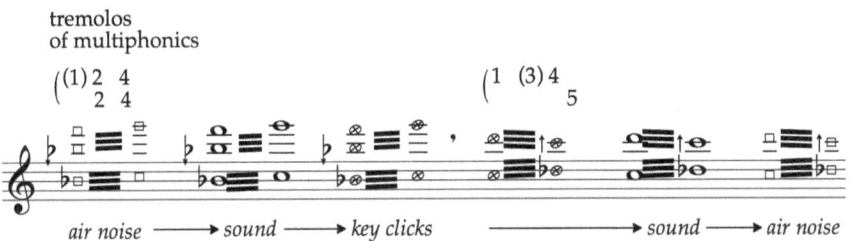

Fig. 21.1 Excerpt from Rakhat-Bi Abdyssagin *Omaggio a Ivan Fedele* per flauto solo (2018), published by Verlag Neue Musik Berlin

This effect was discovered by Pierre-Yves Artaud and is actually a tremolo performed between two multiphonics. When this fast tremolo is performed between given multiphonics (with the indicated fingerings and following sonoristic transition between air noise, sound and key clicks), it sounds not as a stable and static tremolo (like piano tremolos usually sound) but as a magic and sorcerous fluctuations and ever-evolving movement. During these fluctuations as a result of tremolo between multiphonics, some notes and sounds may appear that are not present in either of the two used multiphonics! And sounds that appear during this tremolo between multiphonics are not usually controllable and cannot be achieved by any other way, they are almost 'non existing' or 'virtual' in any other performing conditions and ephemeral in their nature, and are also not present in any of the two multiphonics when they are separated. But when these two multiphonics interact (as a tremolo under sonoristic transition), these sounds appear! It is very similar to quantum entanglement when certain phenomena can occur only when particles are entangled, and when 'the information one can gain by jointly observing the systems is *more* than the information obtained by observing each system separately' (Marletto 2021, p. 126).

Like there, during this tremolo it is possible to gain more information (more notes) than when observing each multiphonic individually. If simply, in this case the interaction of two multiphonics produces more information (tones/sounds) than both multiphonics can produce separately and independently. If these two multiphonics sound independently, the sonoristic information contained in them both (combined) would be less than when they are interacting with each other (tremolo in a given case).

Considering P. Veale's multiphonics for oboe, some of them contain notes that do not normally exist in oboe under any other conditions, so you cannot play them separately, but they are present in a group/combination/cluster and sound together with other notes when a performer plays specific multiphonic.

For example, in my composition *Shimmering* for oboe solo (2020, 2022) I used the following multiphonic (Fig. 21.2) of Peter Veale.

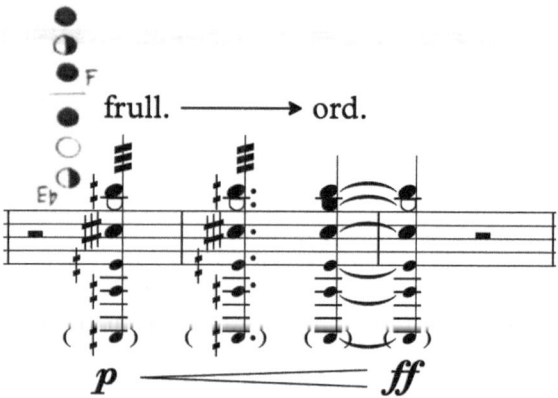

Fig. 21.2 Excerpt from Rakhat-Bi Abdyssagin *Shimmering* for oboe solo (2020) published by Verlag Neue Musik Berlin

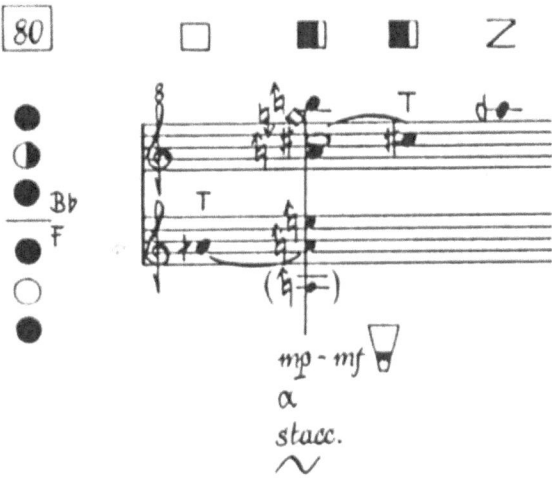

Fig. 21.3 Oboe multiphonic №80 from P. Veale's book, published by Bärenreiter

This multiphonic should sound one octave higher than notated. The low note in brackets (low A quarter-tone sharp which sounds octave higher than written) does not normally exist in oboe; it is usually impossible to play this tone separately. But when a performer plays given multiphonic with the indicated fingering, this low A sounds as part of an inseparable cluster of tones. This effect can be metaphorically linked with entangled states in quantum physics. My piece *Shimmering* for oboe solo was premiered by Peter Veale in Princess Galyani Vadhana Institute of Music, Bangkok, Thailand on 2nd February 2023.

And there are many more such examples for oboe. For example, the multiphonic №80 (Fig. 21.3) from Peter Veale's book (1994, p. 84) is quite similar in structure to the one used in my *Shimmering* described above, and here also the lowest note (indicated in brackets) does not exist in oboe separately, but appears when the whole multiphonic is performed. Unlike the previous multiphonic, this one under №80 is notated in such a way that part on the lower staff sounds as written, and the part on the upper staff sounds an octave higher. This applies also to the following multiphonics.

Even greater examples of such phenomenon can be witnessed in Peter Veale's oboe multiphonics (Fig. 21.4) №68, 69, 74, 96 (1994, pp. 83–84, 86). In all these oboe multiphonics the lowest note (in brackets) absolutely does not exist on oboe separately. These low notes (in brackets) are far beyond the standard range of oboe, and cannot be performed under any normal conditions. But each low note appears when the relevant multiphonic is performed.

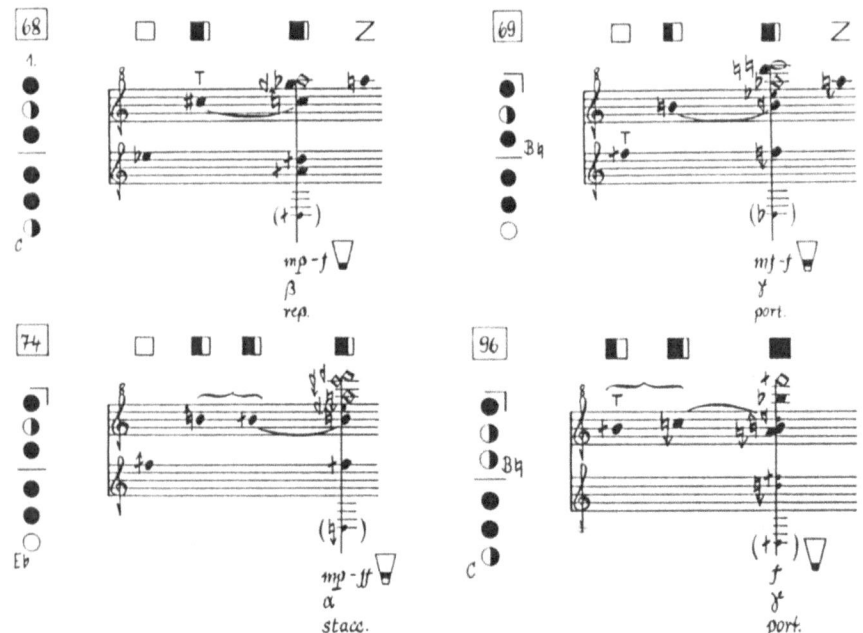

Fig. 21.4 Oboe multiphonics №68, 69, 74, 96 from P. Veale's book, published by Bärenreiter

Therefore this effect can be metaphorically connected with quantum entanglement in physics, as in the reviewed case globally (multiphonics can be characterised as an entangled state of notes) there is more information than locally, so when the whole multiphonic is performed some notes appear that cannot be performed separately (as in a separate form they do not exist on the instrument, and can appear only as part of multiphonic).

Another intriguing example is Peter Veale's oboe multiphonic №72 (1994, p. 83), where the arrow is indicated under the lowest note. This arrow means that while performing this multiphonic there will be low sounds present. These low sounds are beyond oboe's range, and—similar to other multiphonics in consideration—cannot be obtained separately. Nevertheless, (unlike in multiphonics given above) in multiphonic №72 (Fig. 21.5) these low sounds are not fixed and always fluctuate during the performance. So under different performing circumstances different low sounds can be produced. And it is not reasonably feasible to control or precisely predict which low notes will appear, and performer has to work with probabilities of low tones. This is why instead of indicating specific low tone the arrow is drawn, which plainly says that during the performance some low notes may appear.

Fig. 21.5 Oboe multiphonic №72 from P. Veale's book, published by Bärenreiter

Other aspect of some of the above-presented multiphonics is that not only low tones (in brackets), but even some of the highest tones are beyond the range of oboe. This is especially true in the case of the above multiphonics №69, 72, 74, 96. Apart from the lowest tone, the highest tones are also normally not performable on oboe under the conventional method of playing.

In general, there are many more multiphonics existing, and the roster present in this research is only a small fraction of existing multiphonics with similar characteristics and acoustic features.

Another compelling interpretation of 'the whole can be greater than the sum of the parts' lies in an understanding of the importance of information contained in correlations. As Richard Jozsa wrote (during our correspondence in October–November 2023): 'for quantum information, "the whole can be greater than the sum of the parts". This is of course true, and it is a key manifestation of the phenomenon of quantum entanglement. The extra information in the whole, for a system comprising a number of parts, can be understood as residing in correlation information, correlations between the states of the parts, and this is not contained in the info of the parts, each given separately [...] This feature is also present in music in a metaphorical sense—e.g. if we have a melody then its musical content is not contained in the state (pitch/length) of each note given separately, but it is a feature of the global whole (composite system) comprising all the notes i.e. in the relations ("correlations") between them—intervals between the notes and harmony shifts occurring and other global musical structure, beyond the pitch/specification of each note separately'.

These are astonishing thoughts. And they are indeed true. In more complex cases of avant-garde music (with multi-layered structure of harmony, frequencies, timbres, rhythms and formal possibilities etc.) the whole paradigm and 'gestalt' of music are exactly the result of interactions and correlations of dimensions and the overall information is contained in established correlations of the parts. Coming to Jozsa's example with melody, currently (in my music-theory research) I am working on a concept of the 'orbital model of a melody' and the 'timbre-texture co-ordinate'. These concepts are aimed at explaining new musical phenomena that are highly present in contemporary music (actually being its highlight) while—in the current form—are mostly absent in classical music.

During our correspondence, Jozsa wondered 'if it would be possible to find a musical metaphor for *nonlocality* properties of entanglement (that are impossible classically) [...] and how one might even metaphorically represent physical locality in musical terms?'.

Considering a musical metaphor for *quantum nonlocality*, a composer will have to act as a creator of this metaphor. For example, one possible way would be to arrange an instrumentation within an ensemble/orchestra in such a way that instruments 'far away from each other' (e.g. in a sense of range, grouping or location in the score, stage) will 'simultaneously' reflect changes in their parts. Like flute initiating *slap* or other percussive techniques, and double bass immediately following the same concept with *pizzicato* etc. Or making some 'nonlocal' correlations between different groups of instruments (brass vs strings, woodwind vs percussions for instance). The correlations can include very traditional polyphonic methods (vastly present in late medieval, renaissance and baroque music) or the newest composing techniques and extended performing methods.

In such a direction, at the very beginning of the piece a composer could create several independent and separate 'entangled pairs of instruments' (within larger ensemble/orchestra) or 'entangled groups of sounds and their properties' which interact with each other and remain entangled throughout all the development and logical evolution of the dramaturgy of the piece. Or sometimes the entangled pairs of instruments may switch partners and then come back.

Partly this approach was implemented in my orchestral composition *The Sacred Universe of Particles*, composed after my visit to the LHC and CERN in 2018b, as well as in numerous of my other music works.

This direction of thought can lead to a metaphoric realisation of quantum teleportation. If we assume the notes to be the objects themselves, while their context (timbre, dynamics, rhythmic structures, sonoristic techniques, texture etc.) to be the 'states of objects', then entangled pairs of instruments will teleport their states to each other without repeating the whole passage or notes. It can be speculated that metaphorically time in music is not linear—when the initial material is repeated, it symbolises going backwards in time. From that perspective, teleportation or entanglement can be metaphorically implemented in using the initial musical material ('theme') simultaneously with the current phase of form to show the information going backwards and forwards in time simultaneously.

There are various possible instances of 'entangled groups of sounds and their properties'. The Italian composer Alessandro Solbiati widely implements the concept of *figura*. *Figura* (Italian term) is usually represented as a concentrated amount of musical information, a group of specific elements carrying musical information. *Figura* can be a group of specific tones, durations and rhythms, timbral structures and other characteristics, united by the common architectonic idea. *Figura* is more powerful than a tone row or even a harmonic field, though less precise with its rules, in the sense that every *figura* may be created according to different rules and thus represent different systems of structuring musical material. *Figura*'s power lies in the fact that it simultaneously exists in multiple musical dimensions—unifying and being an overarching force over several quasi-independent domains of musical expression, be it pitches/tones, durations/rhythms, timbral-textural relations, orchestral structures etc. *Figura* is a powerful tool since it also accounts simultaneously for structural interrelations, syntax and semantics of music composition. It is exceptionally relevant in orchestral writing. Solbiati's *Sinfonia Seconda* for orchestra (2005) is an example of how multi-dimensional concept of *figura* is implemented in large-scale orchestral work (Fig. 21.6). In this symphony different *figure* (in plural) largely predetermine the dramaturgy, orchestration and all other possible relations (both semantic and syntactic) present in this work. Commonly, *figura*'s essence cannot be found in any of its parts separately, while the interrelations between the parts actually create and construct the essence of a *figura*. As already discussed with quantum information science and quantum entanglement, musical *figura* is a phenomenon not present in any of its parts (even if simply combined together), but existing because of the structure and system of relations between the parts.

Fig. 21.6 Excerpt from the score of A. Solbiati *Sinfonia Seconda*, courtesy of SZ Sugar

References

Abdyssagin, R.-B. 2018a. *Omaggio a Ivan Fedele* per flauto solo. Berlin: Verlag Neue Musik.

Abdyssagin, R.-B. 2018b. *The Sacred Universe of Particles* for Symphony Orchestra.

Abdyssagin, R.-B. 2020. *Shimmering* for oboe solo. Berlin: Verlag Neue Musik.

Abdyssagin, R.-B. 2022. *Selected Solo Works* (Anthology) (2012–2021). Berlin: Verlag Neue Musik.

Artaud, P.-Y. 1995. *Flûtes au présent/Present Day Flutes*. Paris: Billaudot.

Barnum, H., C.M. Caves, C.A. Fuchs, R. Jozsa, and B. Schumacher. 1996. Noncommuting Mixed States Cannot Be Broadcast. *Physical Review Letters* 76 (15): 2818–2821.

Bennett, C.H., G. Brassard, C. Crépeau, R. Jozsa, A. Peres, and W.K. Wootters. 1993. Teleporting an Unknown Quantum State via Dual Classical and Einstein-Podolsky-Rosen Channels. *Physical Review Letters* 70 (13): 1895–1899.

Bohr, N. 1935. Can Quantum-Mechanical Description of Physical Reality be Considered Complete? *Physical Review* 48 (8): 696–702.

Deutsch, D. 1985. Quantum Theory, the Church-Turing Principle and the Universal Quantum Computer. *Proceedings of the Royal Society A* 400 (1818): 97–117.

Deutsch, D., and R. Jozsa. 1992. Rapid Solutions of Problems by Quantum Computation. *Proceedings of the Royal Society of London A* 439 (1907): 553–558.

Dieks, D. 1982. Communication by EPR Devices. *Physics Letters a.* 92 (6): 271–272.

Einstein, A., B. Podolsky, and N. Rosen. 1935. Can Quantum-Mechanical Description of Physical Reality Be Considered Complete? *Physical Review* 47 (10): 777–780.

Ekert, A.K. 1991. Quantum Cryptography Based on Bell's Theorem. *Physical Review Letters. American Physical Society* 67 (6): 661–663.

Gallois, P. 2009. *The Techniques of Bassoon Playing/Die Spieltechnik des Fagotts/La technique du basson*. Kassel: Bärenreiter-Verlag.

Kientzy, D. 1989. *Les sons multiples aux saxophones: Pour saxophones sopranino, soprano, alto, ténor & baryton*. Paris: Éditions Salabert.

Marletto, C. 2021. *The Science of Can and Can't*. Penguin Books.

von Neumann, J. 1932. *Mathematische Grundlagen der Quantenmechanik*. Berlin: Springer Verlag.

Rae, A. 2018. *Quantum Physics: Illusion or Reality?*, 2nd ed. Cambridge: Cambridge University Press.

Rehfeldt, P. 2003. *New Directions for Clarinet*. Lanham, Maryland and Oxford: The Scarecrow Press Inc.

Schrödinger, E. 1935a. Discussion of Probability Relations Between Separated Systems. *Mathematical Proceedings of the Cambridge Philosophical Society* 31 (4): 555–563.

Schrödinger, E. 1935b. Die gegenwärtige Situation in der Quantenmechanik. *Naturwissenschaften* 23 (48): 807–812.

Schrödinger, E. 1936. Probability Relations Between Separated Systems. *Mathematical Proceedings of the Cambridge Philosophical Society* 32 (3): 446–452.

Shor, P.W. 1994. Algorithms for Quantum Computation: Discrete Logarithms and Factoring, pp. 124–134. *Proceedings 35th Annual Symposium on Foundations of Computer Science*. IEEE Computer Society Press.

Solbiati, A. 2005. *Sinfonia Seconda* for Orchestra. Milano: Edizioni Suvini Zerboni.

Timpson, C.G. 2013. *Quantum Information Theory and the Foundations of Quantum Mechanics*. Oxford: Oxford University Press.

Veale, P., C.S. Mahnkopf, W. Motz, and T. Hummel. 1994. *Die Spieltechnik der Oboe/The Techniques of Oboe Playing/La technique du hautbois*. Kassel: Bärenreiter-Verlag.

Watts, S. 2015. *Spectral Immersions: A Comprehensive Guide to the Theory and Practice of Bass Clarinet Multiphonics*. Belgium: Metropolis Music Publishers.

Weiss, M., and G. Netti. 2010. *The Techniques of Saxophone Playing/Die Spieltechnik des Saxophons*. Kassel: Bärenreiter-Verlag.

Wootters, W.K., and W. Zurek. 1982. A Single Quantum Cannot be Cloned. *Nature* 299 (5886): 802–803.

Part IV

Epilogue

22

Epilogue

The purpose of this book is to open the gates to the vast field of possible interconnections and correlations between science and art in general, and quantum mechanics and avant-garde music in particular.

Many people think that science is objective while art is subjective. However, this is not the case. Natural science is not the nature itself, it is our view of nature, the questions we ask about nature and our methods of finding answers. Our questions and methods of finding answers ultimately depend on the development of the human mind, and this development is a result of the overall height of civilisation.

While people perceive art to be subjective or arbitrary, the emergence of style or technique is an objective process driven by global processes that shape the relevant level of evolution of the human mind. Every change in the consciousness of civilisation triggers global changes and shifts of paradigm, and these changes are first perceived by music – the art which reaches the foremost heights of abstraction and the greatest degree of generalisation.

The inception of a style of art or a technique in music is a complex overarching process involving great minds, working quasi independently but essentially intertwined with each other, unfolding mind-structures over generations. Techniques of music composition share many similarities and spiritual connections with mathematical structures. The techniques of composition represent a reflection of the abstract generalised mind-structures of humankind in music.

As I have displayed in this book, the birth of new music was not an accident, but a truly inevitable event! It was impossible for avant-garde music not to be born, especially when all areas of human activity were undergoing

© The Author(s), under exclusive license to Springer Nature
Switzerland AG 2024
R.-B. Abdyssagin, *Quantum Mechanics and Avant-Garde Music*,
https://doi.org/10.1007/978-3-031-63161-0_22

dramatic revolution and evolution. Science is a result of interaction between us and nature, and art is a result of interaction between human mind and the world.

As Heisenberg wrote, 'the two processes, that of science and that of art, are not very different. Both science and art form in the course of the centuries a human language by which we can speak about the more remote parts of reality' (2000, p. 66). Therefore, science and art are different languages in which humanity speaks about the universe.

Reference

Heisenberg, W. 2000. *Physics and Philosophy*. Penguin Classics.

Index